NUCLEAR MATERIALS AND DISASTER RESEARCH

BIOLOGICAL RESPONSES, MONITORING AND PROTECTION FROM RADIATION EXPOSURE

NUCLEAR MATERIALS AND DISASTER RESEARCH

Additional books in this series can be found on Nova's website under the Series tab.

Additional e-books in this series can be found on Nova's website under the e-book tab.

NUCLEAR MATERIALS AND DISASTER RESEARCH

BIOLOGICAL RESPONSES, MONITORING AND PROTECTION FROM RADIATION EXPOSURE

K. P. MISHRA, PH.D.
EDITOR

nova publishers
New York

Library of Congress Cataloging-in-Publication Data

Biological responses, monitoring and protection from radiation exposure / editors: K.P. Mishra, Ph.D. (Vice Chancellor, Nehru Gram Bharti University, Allahabad, India).
pages cm. -- (Nuclear materials and disaster research)
Includes index.
ISBN 978-1-63463-852-4 (hardcover)
1. Radiation--Safety measures. 2. Radiation--Physiological effect. 3. Radiation--Health aspects. 4. Radiation--Toxicology. 5. Radiation injuries--Prevention. I. Mishra, K. P., editor.
RA1231.R2B447 2015
363.17'99--dc23
2014048432

Published by Nova Science Publishers, Inc. † New York

Contents

PREFACE

This book deals with urgent and timely issues related to radiation health effects and protection that are examined by both young researchers as well as experts. The book is organized into three major sections: biological responses, population monitoring and approaches to protection from radiation exposure. Contributors have provided state of the art research in their respective chapters. Radiation action produces damage to multiple targets in the exposed cells or human body and understanding of molecular mechanisms of the underlying processes becomes central to the monitoring of effects and health consequences of radiation exposure. Many experts have highlighted the outcome of epidemiological studies on human populations in high background radiation areas in different locations around the world as well as consequences and scopes for mitigating radiation health effects after radiation accidents such as Chernobyl in Ukraine and the Fukushima Daiichi Accident in Japan. This book also provides important direction for treatment of radiation for exposed victims. In the concluding chapters, contributors have provided new approaches for protection against ionizing radiation exposure.

This book contains rich content on basic aspects of radiation induced cellular response which may give deeper insight to beginners in research, teaching, industry and regulatory authorities for basic understanding of radiobiological processes and molecular mechanisms. The book will prove an authentic reference source for updates in radiation science. It is hoped that students, teachers, experts, safety officers, regulatory officials and policy-makers will find the book handy for gaining a broad view of radiation damage to biological systems, monitoring health consequences and for new approaches in developing effective protection against radiation exposure.

Chapter 1 - Emphasis is placed in this mini-review on recent developments concerning the identification and accurate high-performance electrospray ionization tandem mass spectrometry (HPLC-ESI-MS/MS) measurement of ionizing radiation-mediated formation of damage to cellular DNA. In the first part of the chapter a critical survey of the main single base lesions, likely part of more complex damage, that are generated upon one radical hit either through hydroxyl radical or one-electron oxidation is reported. In addition evidence for the formation of more complex DNA modifications that include tandem base lesions, purine 5',8-cyclo-2'-deoxyribonucleoside, intra- and interstrand cross-links as still the result of a unique radical event is given. The second main section of the contribution deals with the formation of double strand breaks and other oxidatively generated clustered damage that constitute the molecular signature of ionizing radiation on cellular DNA. The effects of secondary low energy electrons are also discussed before the presentation of a few perspectives.

Chapter 2 - The recent advances in the field of stem cell research and regenerative medicine introduce a new and very promising modality in the treatment of degenerative diseases. The findings open a new field of research for the treatment of radiation-induced normal tissue lesions. Normal tissue toxicity is a common side effect of radiotherapy treatment of cancer patients. It is also a major concern in populations subjected to accidental exposure to ionising radiation. At present, there is not a standard protocol for the treatment of radiation-induced normal tissue lesions. Stem cell transplantation or administration of microparticles extracted from stem cells appears to be a promising modality for the treatment of radiation lesions.

Chapter 3 - Biological effects of radiation are not only limited to cells exposed to radiation, but radiation damage gets manifested to bystander cells, which have not directly received radiation. This phenomenon of radiation induced bystander effect (RIBE) has been mainly studied in normal cells with relevance to radiation risk assessment. However, bystander effect pertaining to cancer radiotherapy is very limited in literature. Irradiated cancer cells may interact with bystander tumor cells or normal cells through involvement of gap junctions or secretion of diffusible factors. Moreover, normal cells receiving relatively lower doses may also interact with irradiated cancer cells. The magnitude and nature of bystander signaling in tumor microenvironment would have significant implication in clinical outcome of cancer radiotherapy.

Chapter 4 - The importance of the dose-rate effect in evaluating the cancer risk of ionizing radiation was shown for a wide dose-rate range experienced in the human environment from 10^{-9} Gy/min at the environmental level to 10^{6}-10^{8} Gy/min at the extremely high A-bomb exposure level. The discrepancy between the observed cancer risk and that extrapolated from the A-bomb survivor data with a currently used factor, DDREF, together with the LNT model indicated an overestimation of the risk with the model. Alternately, a non-tumor dose, defined as the highest doses at which no significant cancer increase is observed above the control level was applied to analyze the dependence of the cancer risk on the dose rate at various exposure conditions, including dose rate, difference between whole body and partial body exposures, and LET. The value 20 for DDREF at the environmental radiation level compared with the A-bomb radiation level for the whole-body low LET exposure was presented.

Chapter 5 - Risk and benefit analysis of radiation have engaged the mind of scientific community from several decades. The basic quantification of possible consequences from radiation exposure were estimated from horrendous consequences of Japanese Atomic Bomb survivors. Ionizing radiations and their exposure to human being have the possibility of producing various degrees of harmful effects depending upon the quality and severity of the radiation. Therefore, investigations using ionizing radiation have to be justified by a clear benefit to the patient. In this article a review of quantitative aspects of exposures emanating from conventional X-rays, Multi Slice CT scanners, Interventional radiology practices and mammography have been summarized. The radiation doses from CT scan study comprising of 20 slices may be equivalent to 75- 150 Chest X-ray dose to a single patients. The authors have evaluated the organ dose and ESD (Entrance Surface Dose) emanating from various CT procedures. Well calibrated thermoluminescence dosimeters (TLD) were utilized for measurement of doses. CT as a modality has found new applications in angiography and is being increasingly used, encompassing up to 40% of all radiographic studies. The mean (± SD) skin doses varied from 728 ± 382, to 128 ± 81, mGy, and the mean effective doses varied from 119.7 ± 50.3, to 9.1 ± 5.5 mSv for some of the most performed angiography procedures. The maximum skin dose was 1.95 Gy, in one of the studies. Mammography is an extremely useful non-invasive imaging technique with inbuilt advantages for the detection of early breast cancer but doses to glandular tissues impose risk of induction of malignancy. The doses received by the patients were found to be dependent on compressed breast thickness, for breast cancer patients (0.8 - 3.8 mGy for 7.9 cm to 3.8 mGy for 11.6 cm) and fibrocystic disease / breast cystic (0.3 m Gy for 5.5 cm to 1.6 mGy for 6.2 cm). Some of the somatic effects of radiation depend on the severity of radiation doses which may manifest as sterility (temporary or permanent), loss of appetite, impairment of vision (cataract), leukemia, blackening of skin, shortening of life span, carcinogenesis etc. The article will elaborate its pertinent details. The short-term risk to patients is radiation-induced skin damage in cardiac angiography, which can result from acute radiation doses of ≥2Gy is a matter of concern and challenge to develop such technologies which are safe and accurate. The extent of the skin injury may not be known for weeks after the procedure. Repeated procedures increase the risk of skin injury, because previous radiation exposure sensitizes the skin. Long term effects include the potential risk of cancer. It is generally accepted that there is probably no low dose "threshold" for inducing cancers, i.e. no amount of radiation should be considered absolutely safe. Current article attempts to summarize exposure in diagnostic practices with possible health consequences.

Chapter 6 - Consensus on the radiation effects in cardiovascular system doesn't exist. More evidence is needed to conclude whether or not radiation exposure has increased the incidence of cardiovascular and cerebrovascular disease and associated mortality. The observed excess of cerebrovascular diseases is linked to those having worked during the first six months and having cumulated doses of more than 150 mSv. Although these results are statistically compatible with those of the atomic bombings survivors study, they could only be regarded as preliminary. Latest hibakusha data suppose increase in mortality. Industrial cohorts' studies have controversial results. According to UNSCEAR. "There are deterministic effects, such as cataracts or cardiovascular disease, for which radiation exposure is not the only known cause. If these effects occur, usually some time after high levels of exposure, and there is no specific marker for radiation exposure having caused them, it is not possible to attribute the effect with certainty to radiation exposure, but only to express a probability that radiation was wholly or partly the cause."

The existing level of knowledge allows considering the contributive role of the immune function changes of the late period after irradiation to the cerebrovascular pathology pathways. Therefore, it is also actually taking into account an interaction between the immune and nervous systems. Encephalopathy ("radiation encephalopathy", "discirculatory encephalopathy"- DEP, "demielinisation" encephalopathy) according to medical statistics is one of the most common disorders of a late period after irradiation and one of the most frequent diagnosis for establishing a causal connection of disease with a radiation factor. At the same time, the nature of this disorder is not identified, and many researchers do not recognize it as an effect of irradiation. In some studies it was depicted as a result of direct brain tissue radiation damage, or as an indirect effect due to damage of vascular circulation. Development of the brain pathology as a result of CVD and vegetative dysfunctions, and the substantial vasculatory changes revealed in encephalopathy patients provide support to a latter assumption. More evidence is needed to conclude whether or not radiation exposure has increased the incidence of cardiovascular and cerebrovascular disease and associated mortality.

Chapter 7 - In March 2011, the East coast area of Tohoku district of Japan was attacked by a huge earthquake and succeeding Tsunami, which caused the accident in Fukushima Daiichi Nuclear Power Plant (F1-NPP) leading to a massive release of radioactive nuclides. Nuclear and Industrial Safety Agency (NISA) estimated the released radioactivities of ^{131}I, ^{134}Cs and ^{137}Cs to be 160 PBq, 18 PBq and 15 PBq, respectively. Here the authors overview, especially focusing on radioiodine, the monitoring system and countermeasures by Japanese government and municipalities and the public exposure and health consequences.

Chapter 8 - The nuclear accidents at Fukushima Daiichi Nuclear Power Station (FDNPS) released a large amount of radionuclides into the environment. After the accident, more than 150,000 people were evacuated. Unfortunately, the emergency evacuation was not conducted as timely and efficiently as it should have been. At least 60 elderly who evacuated from hospitals and nursing homes in the area within 20km from FDNPS died by the end of March 2011. During the Fukushima nuclear crisis, a primary concern was the internal exposure of the thyroid gland to radioactive iodine. Stable iodine was not administered by most residents. No extensive effort to measure individual thyroid doses was made, either. By the end of April 2011, when ^{131}I decreased to almost undetectable levels due to its short half-life (=8days), only 1,200 or so residents underwent direct thyroid dosimetry. Another concern was the long-term exposure to radionuclides with relatively long half-lives, such as ^{137}Cs and ^{134}Cs. Fortunately, various surveys have shown that the cumulative radiation doses of residents from those radionuclides will be low. Therefore, the increased risks of malignancies and genetic disorders will be too small to be detected with statistical significance even if the linear non-threshold model is correct, and will probably be buried in the noise of other risk factors, including lifestyles. However, a possible exception is thyroid cancer risk among children as pointed out by the UNSCEAR 2013 Report. It is necessary to examine its risk at least for the next 10 years. The largest health risks after the accident were those related to psycho-social factors rather than the possible direct biological effects of radiation. A lesson from the Fukushima accident is that the people's perception of risk hinges on the trustworthiness of the government, and is not directly linked to the real magnitude of health risk. When the government's credibility is questioned, the hyperinflation of perceived health risk takes place.

Chapter 9 - Some inhabited areas in Ramsar, Iran are known for extraordinary levels of natural background radiation. People in these high background radiation areas (HBRAs) are continuously exposed to high levels of indoor and outdoor gamma radiation as well as high concentrations of indoor Rn-222. In spite of current controversies, some of the studies performed so far have shown increase in the frequency of chromosome aberrations in the lymphocytes of residents of HBRAs. Howevere, no repeatable detrimental effects caused by high levels of natural radiation in Ramsar have been detected so far. Furthermore, the issue of carcinogenesis of high levels of natural radiation in these areas is not proven yet. As the indoor Rn-222 level in Ramsar reaches 31 kBq m^{-3}, Mortazavi et al. previously collected the overall cancer mortality, lung cancer mortality and neonatal death rate in different districts of Ramsar. This study surprisingly showed that the highest recorded rate of lung cancer mortality was in an area with normal levels of indoor radon, while the lowest lung cancer mortality rate was in another area where the highest concentrations of radon in the dwellings were found. Recently, an expert team supervised by Mortazavi studied the tumor markers in the inhabitants of HBRAs of Ramsar. In this study, the serum levels of PSA, CA15.3, CA125, Cyfra21-1, CEA, CA19.9, AFP and Tag72 tumor markers in 35 residents of HBRAs and 35 individuals from a normal background radiation area (NBRA) were measured. Furthermore, the level of background gamma radiation and indoor radon concentration in the houses of all participants were measured. The findings of this study did not show any significant difference between the mean levels of PSA, CA15.3, CA125, CA19.9 and AFP of the residents of HBRAs and NBRAs. However, the levels Cyfra21, CEA and Tag72 in HBRA group were significantly higher compared to those of NBRAs. On the other hand, a statistically significant correlation between the external gamma radiation dose or indoor radon level and the concentration of CEA, Cyfra-21 and TAG 72 biomarkers was observed. As the mean external exposure (indoor and outdoor) in HBRAs of Ramsar is about 6 mSv y^{-1} for a population of about 2000 residents, to obtain statistically significant data in epidemiological studies in the mSv range, large sample sizes are needed. Therefore, to overcome the limitations of sample size in epidemiological studies, the authors believe that studies with relevant tumor markers can shed some light on the dark corners of the challenging issue of cancer incidence in high background radiation areas.

Chapter 10 - Gamma and electron beam radiation would be considered as an attractive method of sterilisation for medical grade polyurethane, due to the radiation resistance provided by the material. However, extremely high irradiation doses can alter the structure and/or function of macromolecules, resulting in oxidation, chain scission and crosslinking. This chapter reviews various topics related to the radiosterilisation of polymers with regards to the technologies used, the effects of radiation on polymers and the mechanisms related to modification and degradation processes that have been reported in the literature. A concise history outlining the development of polyurethane materials is explored; while the behaviour of this material under radiation conditions will be discussed. The importance of environmental conditions and in particular the temperature of the polyurethane material during radiation exposure is highlighted. A detailed account of the behaviour of the material in terms of the thermal, structural, mechanical and surface characteristics will be analysed using experimental data. While it is clear following critical appraisal that irradiated virgin polyurethane will not suffice as an acceptable material for medical device applications, there is methods that yield a high probability for the successful stabilisation of polyurethane material to reduce/eliminate radiation degradation.

Chapter 11 - Survival in radiation induced bone marrow (BM) failure depends on the dose of radiation received and the intensity of supportive care which can protect from otherwise lethal infection and give surviving stem cells a chance to expand. In cases of limited hematopoietic recovery potential, hematopoietic stem cell (HSC) transplantation is the only curative treatment option. The acute radiation syndrome (ARS) similarly to myelosuppression after total body irradiation (TBI) is not restricted to the BM function but can lead to development of multiple organ dysfunctions. Damage of the whole organism is related to a systemic inflammatory response caused by a release of inflammatory cytokines. Management of ARS therefore relies on tissue damage repair processes that might be supported by therapies directed at mitigation of inflammation.

The authors have shown previously the regenerative and immunomodulating effects of BM-derived mesenchymal stromal cells (MSC) e.g. in acute kidney injury. Furthermore, MSC augment hematopoietic recovery after chemo- or radiotherapy transplanted in conjunction with hematopoietic stem cells and are effective in the treatment of steroid resistant graft versus host disease. Therefore the authors tested the potency of MSC as a treatment option after severe radiation exposure.

As a first step, human MSC were subjected to in vitro differentiation into hematopoietic and endothelial lineages. In a second approach, mouse MSC were isolated from BM of mice and i.v. transplanted into lethally irradiated recipients. For long-term observation, leukocyte and thrombocyte counts were determined regularly. After 7 months, all available tissues were investigated for the presence of donor cells. Short-term detection of donor cells was done using quantitative PCR. BM from HSC- and MSC-transplanted animals at day 21 was investigated for differential gene expression.

In vitro, the authors showed aquisition of hematopoietic and endothelial characteristics, albeit to a rather low degree. *In vivo*, the transplantation of mouse MSC led to normalization of white blood cells within 4 weeks. Quantitative evaluation of donor cells in recipients revealed a fast disappearance of donor MSC. Although donor MSC did not home to the BM, the gene expression profile in BM changed significantly. Genes involved in cell cycle, oxidative stress protection, anti-inflammation and detoxification showed upregulation in the BM of MSC-transplanted animals whereas genes responsive for inflammation, protein degradation, adhesion/matrix formation, lipid synthesis and lymphoid development were downregulated, emphasizing the beneficial role of MSC in endogenous hematopoietic reconstitution.

In conclusion, the authors showed a complete reconstitution of the blood system in lethally irradiated mice through transplantation of MSC even without HSC support. Because no homing of donor MSC to the BM was evident they concluded that MSC help to recover from lethal myelosuppression and salvage endogenous HSC. This conclusion was supported by chimerism analysis showing no differentiation of donor cells into HSC *in vivo*. Instead, the milieu in the BM is changed towards a survival-promoting environment counteracting deleterious effects provoked by irradiation. Potential mechanisms of this highly effective trophic mechanism of MSC and its perspectives will be discussed.

Chapter 12 - Radiation exposure from events like accidents in hospitals, nuclear power plant catastrophe such as Chernobyl, Fukushima; nuclear weapon testing fallout from cold war era and use of atom bomb in war e.g. dropping of atom bomb on Hiroshima and Nagasaki in World War II poses threat to human health. In past decades, ionizing radiation has been increasingly used in research, medicine, agriculture and industry. Undesirable effects on health in many of these applications continue to be a matter of concern. Exposure of living cells to radiation generates reactive oxygen species (ROS) inside cells and thereby induces oxidative stress (OS) leading to various OS related diseases. Apart from ROS generation, radiation is known to cause DNA strand breaks, chromosomal aberrations, mutations and triggering apoptosis in the normal as well as in tumor cells. Extensive research has shown the ability of herbal drugs to protect normal cells from the ionizing radiation. Results from *in vitro* and *in vivo* studies have also shown that some herbal drugs have ability to act as anticancer drug in cancer radiotherapy but effectively protect normal cells against radiation damage. This review aims to discuss the mechanisms of action of herbal drugs in reduction of oxidative stress, repair of DNA damage, prevention of lipid peroxidation (LP) and activation of anti apoptotic processes by radiation in normal and tumor cells. A few examples are given concerning the mechanism of action of herbal formulations which point to potential of their practical applications in radioprotection and radiotherapy.

Chapter 13 - Exposure of cells to ionizing radiation IR) induces generation of reactive oxygen species (ROS) which may damage vital cellular molecules if not detoxified. Many herbs are known to protect cells from deleterious effects of induced ROS due to their antioxidant (AOs) properties. It is observed that endogenous antioxidants like SOD, GSH and GPX have potential to neutralize the stress produced due to ROS resulting in promotion of cellular survival. This paper describes the prospects of searching potent herbal antioxidants which possess radioprotective capability and their influence on initial and long term radiation damage to cells after radiation exposure. Furthermore, they can be regularly consumed as dietary component. Although the exact mechanism is unclear but it seems possible that these AOs recognize the cytosolic redox status of cells which is known to be different in normal and stressed cells. Another possibility is that AOs up regulate endogenous defense machinery reflected in enhanced intracellular SOD or GSH. Many AOs from herbal sources have been reported to produce protective effect in tissues exposed to radiation. The molecular mechanism is reported to involve their ability to down regulate ROS below toxic levels and consequent protection against radiation.

Chapter 14 - Ionizing radiation (IR) can cause damage to DNA molecules, both through direct ionization and through induction of oxidative stress. Thus the resistance and the rescue to radiation-induced damage are mostly associated with the efficacy of DNA repair system, which involves different types of enzymes and signal molecules, many of which are not yet identified. These enzymes represent target for molecules that, modulating their activity, may increase or decrease the efficiency of radiation-induced DNA repair and consequently cell viability. Topoisomerases (Tops) are ubiquitous and essential enzymes that solve the DNA topological problems deriving from replication, transcription, chromatin assembly, recombination, and chromosome segregation by introducing transient breaks into the helix. Human Topoisomerase IB (Top IB) is also of significant medical interest being the selective target of antitumor drugs such as camptothecin and its water-soluble derivatives, widely used for treating cancers sometimes in association with radiotherapy. There are several evidences that Top IB is involved in DNA repair, although contradicting results have been reported. The role of Top IB in repairing DNA damage induced by X rays was studied on two human cancer cell lines (i.e. colorectal carcinoma - HCT 116 - and adenocarcinoma - HT 29) evaluating the relation between Top IB activity, and the level of DNA damage. After IR exposure, the DNA repair is correlated to an increase of Top IB activity but not of its synthesis. Based on these outcomes and on the available literature, the authors figure that the radiation exposure induces a post-translational modification of the enzyme modulating its activity. The authors' future prospective is to identify these post-translational modifications, and also to find out relation between Top IB and other repair proteins. These results provide a new insight into the mechanisms for genomic integrity maintenance, and could also contribute to new strategies in cancer treatment.

In: Biological Responses, Monitoring and Protection … ISBN: 978-1-63463-852-4
Editor: K. P. Mishra

Chapter 1

RADIATION-INDUCED DAMAGE TO CELLULAR DNA: RECENT ASPECTS AND PERSPECTIVES

***Jean Cadet*[1,2,3] *and J. Richard Wagner*[3]**

[1]Univ. Grenoble Alpes, INAC, Grenoble, France

[2]CEA, INAC, Grenoble, France

[3]Département de Médecine Nucléaire et Radiobiologie, Faculté de Médecine de des Sciences de la Santé, Université de Sherbrooke, Sherbrooke, Québec, Canada

ABSTRACT

Emphasis is placed in this mini-review on recent developments concerning the identification and accurate high-performance electrospray ionization tandem mass spectrometry (HPLC-ESI-MS/MS) measurement of ionizing radiation-mediated formation of damage to cellular DNA. In the first part of the chapter a critical survey of the main single base lesions, likely part of more complex damage, that are generated upon one radical hit either through hydroxyl radical or one-electron oxidation is reported. In addition evidence for the formation of more complex DNA modifications that include tandem base lesions, purine 5',8-cyclo-2'-deoxyribonucleoside, intra- and interstrand cross-links as still the result of a unique radical event is given. The second main section of the contribution deals with the formation of double strand breaks and other oxidatively generated clustered damage that constitute the molecular signature of ionizing radiation on cellular DNA. The effects of secondary low energy electrons are also discussed before the presentation of a few perspectives.

Keywords: Ionizing radiation, base lesions, hydroxyl radical, one-electron oxidation, cyclonucleosides, oxidized guanine, tandem lesion

1. INTRODUCTION

Considerable efforts have been devoted during the last five decades to the identification of the main radiation-induced decomposition products of both the nucleobases and the sugar moiety of nucleic acids. In that respect one of the major initial discoveries at the end of the 50's was the peroxidation of pyrimidine bases [1] and more specifically of thymine [2] by hydroxyl radical ($^{\bullet}OH$) generated by x-radiolysis of water molecules. This has provided a strong impetus to the development of model studies aimed at characterizing radiation-induced degradation products of thymine and cytosine [3, 4]. At the same period attempts were made to search for 5,6-dihydroxy-5,6-dihydrothymine, the so-called thymine glycol that is a stable degradation product of 6-(5)-hydroperoxy-5-(6)-hydroxy-5,6-dihydroperoxides [5] in cellular DNA. Unfortunately the two assays designed for this purpose that involved either a mild reduction [6, 7] or alkaline degradation [8, 9] of thymine glycol had led to misleading and inconsistent data due to the occurrence of major drawbacks as outlined in an earlier review article [10] and further discussed recently [11]. Subsequently, two chromatographic methods were developed by two independent groups and they immediately received a great deal of attention. The association of high performance liquid chromatography with electrochemical detection (HPLC-ECD) was successfully applied for the measurement of 8-oxo-7,8-dihydro-2'-deoxyguanosine (8-oxodG) [12], following its discovery a couple of years before [13] in the DNA of X-irradiated rats [14]. The second assay that involved the mass spectrometry detection of silylated derivatives of oxidized nucleobases at the output of gas chromatographic column (GC-MS) was proposed to measure more than 20 radiation-induced lesions in addition to 8-oxo-7,8-dihydroguanine [15, 16]. However it was subsequently established by several groups that the reported data provided by the GC-MS method suffered from major drawbacks and were overestimated by 2 to 3 orders of magnitude [17-19]. Then the availability at the end of 90's of the HPLC-electrospray ionization tandem mass spectrometry (ESI-MS/MS) [20, 21] and the design of suitable conditions of DNA extraction [22-24] have allowed to accurately single out the radiation-induced formation of several oxidatively pyrimidine and purine base modifications in cellular DNA. It may be added that major progress has been achieved during the last two decades in the elucidation of oxidative degradation pathways of isolated DNA and model compounds including 2'-deoxyribonucleosides and short oligonucleotides (for recent reviews see 25-30). Altogether this had led to a comprehensive understanding of the mechanisms of action of ionizing radiation on cellular DNA in terms of indirect and direct effects that involve mostly $^{\bullet}OH$ and one-electron oxidation reactions respectively. Therefore emphasis is placed in this short review on recent aspects of the radiation formation of single and more complex oxidatively generated modifications in cellular DNA. Clustered lesions initially called "locally multiply damaged sites" (LMDS) [31, 32] that are formed as the result of several simultaneous radical hits [33-35] are considered as the molecular signatures of ionizing radiation. These aspects are discussed in the second part of the chapter and a few perspectives and expected developments are presented in the conclusions.

2. Hydroxyl Radical Induced DNA Damage

As predicted from Monte-Carlo calculations [36-38] most of modified bases and single strand breaks (SSBs) are part of oxidatively generated clustered lesions (OGCLs), each of them being almost unique with numerous possible combinations while their identification remains a challenging analytical issue. It may be added that structural information on these complex types of damage is lost during either acidic hydrolysis or enzymatic digestion prior to chromatographic analysis that is now mostly achieved using HPLC-ESI-MS/MS. In most cases the base modifications are measured as 2'-deoxyribonucleosides with the exception of 2 -diamino-4-hydroxy-5-formamidopyrimidine (FapyGua) and 4,6-diamino-5-formamidopyrimidine (FapyAde) [39] due to the high hydrolytic lability of the *N*-glycosidic bond of the corresponding nucleosides [40, 41]. As further discussed below, damage mediated by ionizing radiation to the bases of cellular DNA can predominantly (80%) accounted for by $^{\bullet}OH$ according to indirect effects while ionization likely contributes to the remaining 20%.

2.1. Single Base Lesions

Among the estimably eighty $^{\bullet}OH$-mediated oxidation products of pyrimidine and purine 2'-deoxyribonucleosides that have been identified in model studies performed in aerated aqueous solutions, only 20 oxidized nucleosides including diastereomers have been detected and their formation quantified so far in the DNA of gamma-irradiated mammalian cells using the accurate HPLC-ESI-MS/MS method. It may be pointed out that relatively high doses of ionizing radiation of between 40 Gy and 2kGy have been used in order to detect a significant increase in the steady-levels of base lesions that arise from oxidative metabolismwith however a small artifactual contribution due to spurious oxidation of overwhelming 2'-deoxyribonucleosides during DNA isolation [42]. Interestingly, it was found using the modified alkaline comet assay in combination with repair DNA *N*-glycosylases [43, 44] that the formation of base damage is linear with the radiation doses from 1 Gy to 10 Gy with a possible extrapolation to higher doses [45]. In the initial studies the four *cis* and *trans* diastereomers of 5,6-dihydroxy-5,6-dihydrothymidine (dT-Gly) that arise from $^{\bullet}OH$ addition at C5 and to a lesser extent at C6 of the thymine moiety were detected in the DNA of human monocyte cells as the main degradation products [46, 47]. In addition, 5-(hydroxymethyl)-2'-deoxyuridine (5-HmdU) and 5-formyl-2'-deoxycytidine (5-FodU) whose formation is rationalized in terms of initial $^{\bullet}OH$-mediated hydrogen abstraction from the methyl group of thymine were shown to be generated [46, 47] (Figure 1). Recently the two diastereomers of 1-(2-deoxy-*β*-D-*erythro*-pentofuranoyl)-5-hydroxy-5-methylhydantoin (Hyd-dT) that are formed through opening of 5,6-bond of peroxide intermediate(s) and subsequent rearrangement [48] have been detected in the DNA of cultured Fischer glioma (F98) cells upon exposure to gamma rays [49]. In the same study, detailed information was gained into several $^{\bullet}OH$-mediated degradation pathways of the cytosine moiety in cellular DNA through the monitoring of four oxidized 2'-deoxyribonucleosides by HPLC-ESI-MS/MS [48]. These include 5-hydroxy-2'-deoxycytidine (5-OHdC) and 5,6-dihydroxy-5,6-dihydro-2'-deoxyuridine (dU-Gly) that arise from dehydration and deamination, respectively, of initially generated 5,6-dihydroxy-5,6-dihydro-2'-deoxycytidine (dC-Gly) [50]. In addition, 3-(2-deoxy-*β*-D-*erythro*-pentofuranoyl)-1-carbamoyl-4,5-dihydroxy-2-oxoimidazolidine (Imid-dC) and 1-(2-deoxy-*β*-D-*erythro*-pentofuranoyl)-5-hydroxyhydantoin (Hyd-dU), two different rearrangement products of the pyrimidine ring [51, 52] were also detected. In a subsequent study information was provided on the radiation-induced formation of several $^{\bullet}OH$ -induced degradation products of 5-methyl-2'-deoxycytidine (5-mdC), a relevant epigenetic mark of gene expression [53], in the DNA of F98 cells [54]. The two main oxidation products of 5-

mdC that have been detected consisted of 5-(hydroxymethyl)-2'-deoxcytidine (5-HmdC) and 5-formyl-2'-deoxycytidine (5-FodC). It may also be emphasized that purine bases are also subject to •OH-mediated reactions in cellular DNA that give rise to the preferential formation of FapyGua with respect to 8-oxodG for guanine, as the result, respectively, of one-electron reduction and one-electron oxidation of initially generated 8-hydroxy-7,8-dihydroguanyl radical [55]. The yield of formation of FapyAde and 8-oxo-7,8-dihydro-2'-deoxyadenosine (8-oxodA), the related adenine degradation products is about 10-fold lower than that of FapyGua and 8-oxodG [46]. This may be explained at least partly by the occurrence of one-electron oxidation reaction of 2'-deoxyguanosine (dG) mediated by pyrimidine peroxyl radicals with vicinal guanine that leads to an enhancement of the formation of 8-oxodG [56] (Figure 2).

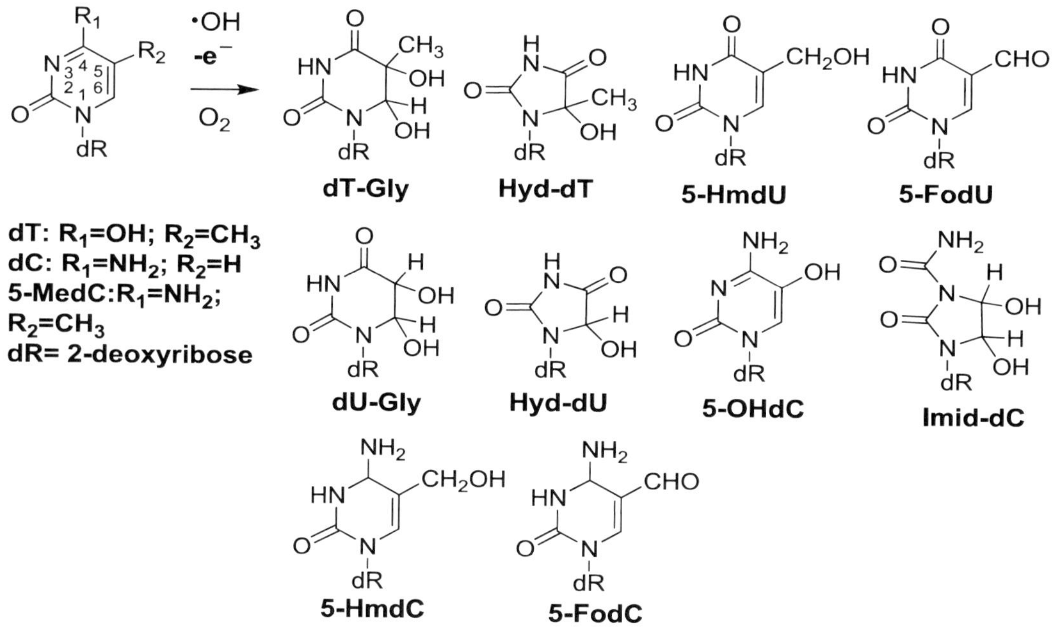

Figure 1. Oxidized pyrimidine bases.

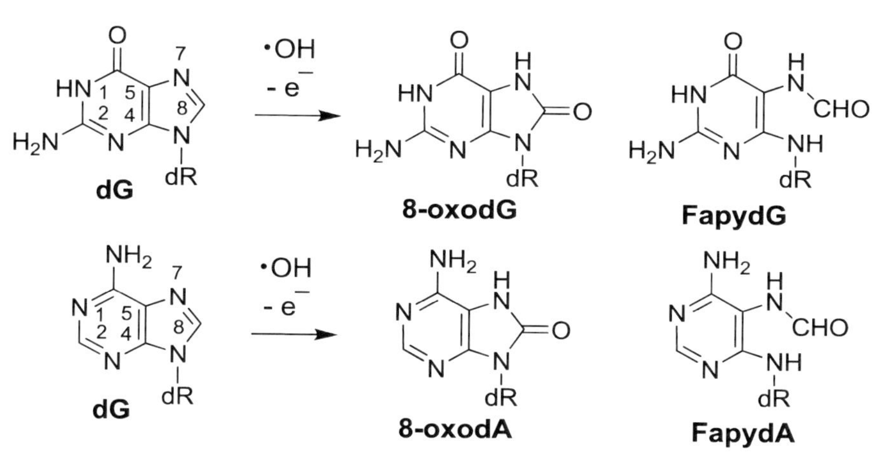

Figure 2. Oxidatively generated purine base modifications.

2.2. Intrastrand Base Cross-Links

Detailed model studies [57-59] that were complemented by theoretical investigations [60, 61] have shown the occurrence of preferential intramolecular addition of 5-(uracilyl)methyl radical to vicinal guanine when located at the 5'-site [58, 59]. In addition the latter exocyclic radical that is derived from thymine upon either •OH-mediated hydrogen abstraction from the methyl group or one-electron oxidation reacts with a much lower efficiency with adenine. Interestingly the formation of G[8-5m]T, the thymine adduct to guanine that is partly prevented by the fast reaction of O_2 with 5-(uracilyl)methyl radical, has been detected by HPLC-MS^3 in the DNA of HeLa-S3 cells after exposure to gamma rays [62]. The interstrand base cross-link is generated in a much lower yield than 5-HmdU and 5-FordU [63], the subsequent final degradation product of the peroxyl radical or related hydroperoxide formed by the addition of O_2 to 5-(uracilyl)methyl radical [48]. A second vicinal base adduct whose formation is rationalized by addition of the 6-hydroxy-5,6-dihydrocytos-5-yl radical to the C8 of 5'-guanine has been shown to be generated upon gamma irradiation of cellular DNA [64]. The resulting addition product is likely to undergo dehydration giving rise to G[8-5]C, a 5,6-unsaturated cytosine containing cross-link (Figure 3).

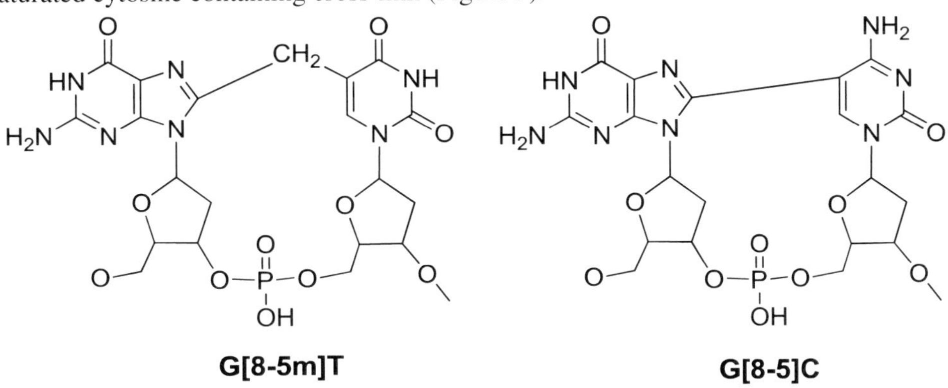

Figure 3. Oxidatively generated tandem base modifications.

2.3. Hydrogen Atom Abstraction from the 2-Deoxyribose Moiety

The radiation-induced formation of single strand breaks (SSBs) is a well documented pathway in cellular DNA. About 1000 strand breaks that may include alkali-labile sites were estimated to be produced per Gy in diploid mammalian cell when the measurement is achieved using the alkaline sucrose sedimentation technique 33, 65, 66]. The generation of SSBs in isolated DNA is mostly accounted for by •OH-mediated hydrogen abstraction from 2-deoxyribose at C3, C4 and C5 [67-69]. In addition oxidized abasic sites that include 2-deoxyribonolactone (dL) and the C4 oxidized residue (C4-AP) are produced through the transient generation of peroxyl radicals that result from the addition of O_2 to C1 and C4 centered radicals respectively [68, 70, 71]. GC-MS and HPLC-MS based methods have been designed to measure dL and C4-AP lesions in isolated and cellular DNA. [72, 73]. In addition as further discussed below intrastrand and interstrand cross-links whose formation is initiated by •OH-induced hydrogen abstraction from the 2-deoxyribose backbone are generated through subsequent reactions with vicinal nucleobases on the same or opposite DNA strand in cellular DNA.

2.3.1. Purine 5',8-Cyclo-2'-Deoxyribonucleosides

The formation of purine 5',8-cyclo-2'-deoxyribonucleosides that arise from the •OH-mediated generation of 2-deoxyribos-5-yl radical was observed from model studies involving 2'-deoxyribonucleosides and nucleotides almost 40 years ago (for a recent review, see 74). Additional mechanistic information on this relevant class of tandem modifications that are substrates for the nucleotide excision repair (NER) pathway [75-77] and not the usual base excision repair (BER) pathway for single oxidized bases [75] has been obtained recently. Thus it was shown the initial formation of 2-deoxyribos-5-yl radical is followed by intramolecular cyclization at C8 of adenine and guanine with a rate constant of $k = (1.6 \pm 0.2) \times 10^5\ s^{-1}$ and $(6.9 \pm 0.8) \times 10^5\ s^{-1}$, respectively [78, 79]. In a subsequent step oxidation by O_2 of the resulting aminyl radical thus formed gives rise to the 5'*R* and 5'*S* diastereomers of either 5',8-cyclo-2'-deoxyadenosine (cyclodA) or 5',8-cyclo-2'-deoxyguanosine (cyclodG), the 5'*R* form being predominant in both cases by 3- to 4-fold with respect to the related 5'*S* compound [80, 81]. GC-MS and HPLC-MS measurements of either cyclodG or cyclodA in cellular DNA upon exposure to low doses of ionizing radiation have led to questionable data with yields of purine cyclo-2'-deoxyribonucleosides higher than those found in previous studies for 8-oxodG [82, 83], one of the predominant •OH-induced oxidized nucleosides [46, 47]. In fact using accurate HPLC-MS/MS only the (5'*R*)-cyclodA was detected in DNA of THP1 monocytes upon exposure to gamma rays at a dose of 2 kGy [80]. The formation efficiency of (5'*R*)-cyclodA is about 1% of that of 8-oxodGuo; this is partly explained by the low •OH induced generation of C5-yl radical within adenine nucleotide which represents less than 3% of overall DNA reactions. Furthermore most of the latter sugar radical is efficiently scavenged by O_2, thus allowing only a small fraction available for cyclodA formation as shown in model studies [80] (Figure 4).

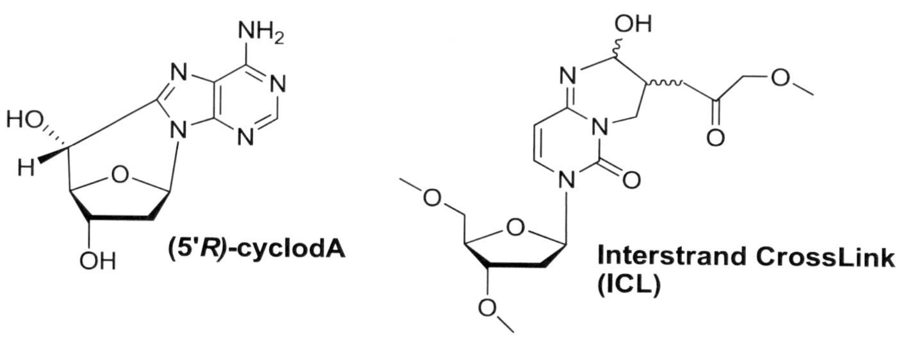

Figure 4. Complex lesions generated by OH-mediate hydrogen abstraction from the 2-deoxyribose moiety.

2.3.2. Interstrand Cross-Link

Evidence has been provided from model studies that have involved the use of HPLC-ESI-MS/MS in neutral ion loss mode that C4-AP generated in one DNA strand is able to react with opposite cytosine to produce intrastrand crosslinks (ICLs). The adducts thus formed in isolated DNA were released as modified 2'-deoxyribonucleosides, detected by HPLC-ESI-

MS/MS and identified on the basis of extensive NMR measurements as the four diastereomers of 6-(2-deoxy-ß-D-*erythro*-pentofuranosyl)-2-hydroxy-3(3-hydroxy-2-oxopropyl)-2,6-dihydroimidazo[1,2-c]-pyrimidin-5(3*H*)-one [84, 85]. Interestingly the four diastereomeric 2'-deoxycytidine (dC) adducts were found to be generated in the DNA of human THP1 monocytes upon either exposure to gamma rays or incubation with bleomycin [85] that selectively leads to H-atom abstraction at C4 of the 2-deoxyribose moiety [86]. It may be added that the mechanism of formation of these ICLs has benefited from model studies involving the site-specific generation of the C4-AP from suitable photo-labile precursors inserted into defined sequences in oligonucleotides [87, 88]. It was shown that the presence of cytosine and to a lesser extent of adenine on the opposite strand in front of C4-AP promotes the conversion of a highly reactive unsaturated keto-aldehyde via a β-elimination reaction with the concomitant formation of a strand break on the 3'-end. In a subsequent step cycloaddition of the latter aldehyde across the 5,6-ethylenic bond of dCyd gives rise to the formation of a ICL in close proximity of a single strand break with a frequency which is about two orders of magnitude lower than that of 8-oxodG. The repair of this complex type of lesions was found to be slow since more than 24 h were required to remove almost quantitatively the modified dC residues in cellular DNA that was treated with bleomycin.

3. One-Electron Oxidation of Nucleobases

As already mentioned one-electron oxidation of the nucleobases and 2-deoxyribose moieties of cellular DNA through the direct effect of ionizing radiation is a minor process with respect to •OH-mediated degradation reactions associated with the indirect effects. A large body of information on the chemical reactions of the base radical cations has been gained from model studies that have mostly involved purine and pyrimidine 2'-deoxyribonucleosides and have been extended to oligonucleotides and isolated DNA [25-27, 29, 30, 53]. For this purpose, type I photosensitizers including mainly riboflavin and 2-methyl-1,4-naphthoquinone have been used as one-electron oxidants of the bases. As an alternative that has received relevant applications in cellular DNA, high intensity nanosecond UVC laser pulses have been shown through a bi-photonic mechanism to ionize purine and pyrimidine nucleobases with a similar efficiency [89, 90].

3.1. Single Base Lesions

UVC laser irradiation of cellular DNA has led to the formation of several base oxidation products that have been detected by HPLC-ESI-MS/MS. In the first study which was performed on TPH1 monocytes, 8-oxodGuo was formed predominantly over three thymidine degradation products including dT-Gly, 5-HmdU and 5-FodU [47]. The generation of these oxidized nucleosides was rationalized in terms of initial formation of guanine and thymine radical cation followed by either hydration or deprotonation and subsequent fast reaction of the resulting carbon centered radical with oxygen [26, 29, 30, 53]. However the relatively high yield of 8-oxodG cannot be only explained by hydration of initially generated guanine radical cation [91, 92] but also through a strong contribution of hole transfer reaction from one-electron oxidized adenine and pyrimidine base residues to somewhat distant guanine base as observed in model studies [93, 94]. In a recent investigation 5-OHdC was found to be generated in a minor amount with respect to 8-oxodG in the DNA of HeLa S3 cells after exposure to intense nanosecond 266 nm Nd:Yag laser pulses [95]. A reasonable mechanism for the formation of 5-OHdC involves initial generation of cytosine radical cation followed by hydration and O_2 addition giving rise to dC-Gly and subsequent dehydration [25].

3.2. Intrastrand Cross-Links

Evidence for the formation of an intrastrand cross-link between guanine and thymine separated by one extra base has been obtained in the DNA of cultured HeLa cells upon UVC laser irradiation [95]. This was achieved after DNA enzymatic hydrolysis and subsequent HPLC-MS/MS analysis. The formation of the G*-T* adduct that constitutes another example of nucleophilic additions to the guanine radical cation as inferred from detailed mechanistic studies [96, 97] is a minor process since it represents about 1% of the yield of 8-oxodG [95] (Figure 5).

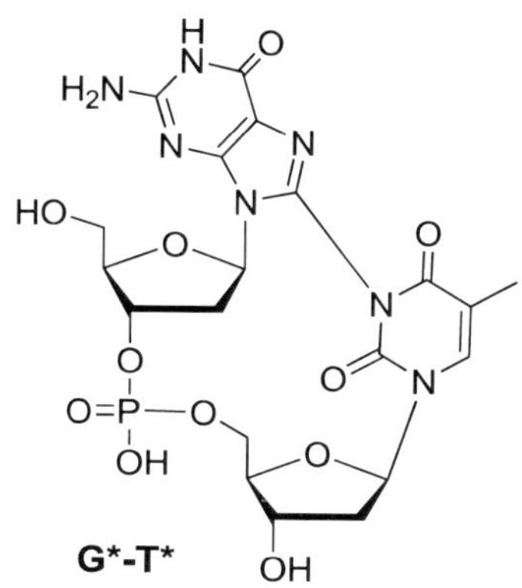

Figure 5. Intrastrand cross-link produced through one-electron of guanine.

4. Damage Induced by Multiple Radical Events

A characteristic feature of ionizing radiation is the multiplicity of radical and excitation events at the site of energy deposit along the photon track giving rise to complex types of DNA damage whose characterization remains a challenging analytical issue.

4.1. Double Strand Breaks

DNA double strand breaks (DSBs), a broad class of clustered DNA damage, consist of two single strand breaks on opposite strands separated by less than 15 base pairs. Various methods including pulsed-field gel electrophoresis [98], neutral comet assay [99] and fluorescence detection of GFP-tagged DNA repair proteins as DSB response markers [100] are available for measuring DSBs in mammalian cells. Immunodetection of gamma-H2AX foci is a very sensitive assay for detecting DSBs [101-103] even if the specificity of the method has been questioned [104]. Relevant information on the spatial distribution and structure of radiation-induced DSBs has been inferred from Monte-Carlo track simulations by taking into consideration the linear transfer parameter of the radiation [36-38, 105-107]. As a striking prediction it was found that the frequency of complex DSBs comprising at least one additional DNA nick or modified base increases up to 70% for 2 MeV particles while the proportion was only 20% for low linear transfer energy (LET) 4.5 keV electrons [36]. Experimental support for the induction of more densely generated DSBs with the increase in LET of incident HZE particles was provided by the observation of a larger frequency of fluorescent γ–H2AX clusters [108]. It has been shown that DSBs generated upon exposure to high LET radiation are more deleterious since their repair involving predominantly the non-homologous end-joining (NHEJ) pathway [109-111] appears to be more problematic [112, 113]. This is likely due to an increase in the complexity of the clustered lesions that may contain additional base lesions and/or oxidized 2-deoxyribose residues at the extremities of the DNA strands.

4.2. Oxidatively Generated Clustered DNA Damage

A second class of radiation-induced complex DNA damage consists of non-DSB clusters that may comprise at least one SSB on one strand together with one or several base lesions and/or apurinic sites on the two complementary DNA strands within one or two DNA helix turns [35, 114]. Attempts have been made using a DNA repair based assay to detect and quantify some of these bi-stranded lesions that contain at least one oxidized pyrimidine base, one modified purine base or an abasic site [115-117]. Surprisingly, the yields of these clustered lesions that could be converted into DSB upon incubation with dedicated DNA repair glycosylases are similar to those of single oxidized bases [118, 119]. These values have to be compared with the yield of radiation-induced DSBs, another class of clustered lesions, that has been estimated to be about 50-fold lower than that base modifications [33]. This apparent inconsistency is strongly suggestive of an artifactual overestimation of the frequency of these non-DSB clustered lesions [120]. It is clear that further work is required to better assess the formation of non-DSB bi-stranded modifications whose measurement remains a highly challenging analytical issue. At present, only Monte Carlo simulations are able to provide estimations on the distribution and composition of the radiation-induced non-DSB clusters [105, 115].

5. Other Radiation-Induced DNA Degradation Pathways

5.1. Low Energy Electron

Numerous species are generated along a high energy radiation track, including secondary electrons, positive and negative ions, and radicals. Among these species, secondary electrons having a most probable energy of about 10 eV are the most abundant of the initially formed species (~ 4 × 10^4 /MeV) [121]. Low energy electrons (LEEs) with energies of 10 eV and less can interact with DNA components by the initial formation of transient molecular ions that subsequently dissociates within a short time frame (10^{-14} s) by a process known as dissociative electron attachment (DEA) [122]. LEEs can also transfer energy to the molecule and then auto-detach leaving the molecule in an excited and possibly dissociative state. The above reactions have been studied for a wide variety of DNA model compounds, including nucleosides, nucleotides, oligonucleotides and purified plasmid DNA [123-136]. The targets are irradiated in the solid state usually under high vacuum at room temperature. After irradiation, the targets are dissolved in solution as a means to identify the chemical modifications by various methods: HPLC, LC-MS/MS and gel electrophoresis. A number of reaction pathways have been described to account for LEE induced DNA damage (< 10 eV). One reaction of LEEs with DNA involves cleavage of the *N*-glycosidic bond by DEA leading to release of a nucleobase anion and a carbon-centered 2-deoxyribose radical. This reaction is demonstrated by the release of nonmodified bases (Thy, Cyt, Ade, Gua), with a preference for the release of Thy and Ade, and for the release of bases at the termini of short oligonucleotides [123, 124]. A second important reaction of LEEs with DNA involves cleavage of the C-O bond giving a carbon-centered radical either at C3 or C5 of the 2-deoxyribose moiety together with a phosphate anion. The corresponding 2',3'-dideoxyribonucleosides resulting from reduction of the carbon centered radicals at C3 of the sugar moiety of T, C, A and G have been identified in isolated and cellular DNA exposed to gamma radiation [132]. The reaction of LEEs (10 eV) with DNA components has also been shown to give DNA base modifications. A major product of the reaction of LEEs with thymidine and thymine containing oligonucleotides (e.g. TpTpT) gives the well-known two-electron reduction product of thymine, 5,6-dihydrothymine [129, 136]. A likely mechanism for the formation of 5,6-dihydrothymine involves the addition of either hydrogen atoms or hydride anions, which are generated by primary reactions of LEE, to the 5,6-double bond of thymine. Lastly, two additional base modifications, 5-hydroxymethyluracil and 5-formyluracil, were identified as products from the bombardment of TpTpT with LEE (10 eV) [135]. Likewise, the formation of these products was shown to take place using an alternative irradiation system in which LEEs with a peak energy of less than 2 eV are generated at the surface of a film of thymidine and tantalum under ambient pressure and temperature [136]. These studies indicate that the reaction of LEEs with thymine leads to the formation of 5-(uracilyl)methyl radicals as the primary precursor of 5-hydroxymethyluracil and 5-formyluracil. A likely route to the formation of 5-(uracilyl)methyl radicals involves cleavage of the C-H bonds at the methyl group of thymine by DEA giving the latter radical with the concomitant formation of a hydride anion (the hydride anion can go on and cause subsequent damage such as 5,6-dihydrothymine). A salient feature of LEE-induced DNA damage is the likely occurrence of two simultaneous bond breaking events in close proximity as a result of multiple losses of energy by a single LEE. For example, an electron of 10 eV may lose 7 eV in an initial interaction with a molecule leading to its excitation, and the electron may detach from the molecule with a lower energy (3 eV) and subsequently induce additional damage in close proximity to the initial damage [128, 130]. This may explain why single strand breaks are readily formed by the reaction of LEEs with energies below 4 eV with plasmid DNA whereas electrons with higher energy are necessary to induce double strand breaks [137].

5.2. Bystander Effect

Aside from immediate radiation-induced damage to cellular DNA generated by the deposition of energy, evidence has been provided more than 20 years ago for the transmission of genomic instability to vicinal cells and their progeny upon α-particle irradiation [138, 139]. These observations have provided a strong impetus to the development of extensive research projects aimed at investigating non-targeted effects associated with the exposure of cells or organisms to ionizing radiations. Radiation-induced bystander effects (RIBE), also called the "abscopal effect" in radiotherapy, appear to be triggered by intercellular communication as inferred by studies with a wide range of cell types and to a lesser extent in vivo [for comprehensive reviews see references¨140-145]. It may be added that these non-targeted effects associated with ionizing radiation exposure have a significance biological incidence for doses lower than 100 mGy. Asalready discussed, DSBs represent highly deleterious lesions that have been proposed to be involved in the "DNA damage response", the likely initiator of the intercellular signaling events leading to RIBE. It was proposed that the signals generated in irradiated cells with a time frame of few minutes [146] could be transmitted to neighboring non-target cells by secreted factors in the medium and/or via connexin 43 (CX43)-mediated gap-junction intercellular communication [147]. It may be pointed out that implication of cytokines, chemokine and growth factors appears to be essential [148]. This is accompanied by changes in the redox status of non-irradiated cells with an exaltation of the production of reactive oxygen species through initial formation of low-reactive superoxide radical anion ($O_2^{\bullet -}$) [149] through cellular metabolism and the induction of cytosolic enzymes [147]. The release of nitric oxide ($^{\bullet}$NO) has been also proposed to be enhanced through activation of nitric oxide synthases (NOS). The increased formation of both $O_2^{\bullet -}$ and $^{\bullet}$NO is known to be associated with inflammation processes that are triggered by the activation of the stress response signaling [150, 151]. Non-targeted effects are characterized by oxidatively generated damage to DNA [147, 152] inferred in particular from the immunofluorescence detection of γ-H2AX foci in vicinal non-irradiated cells [153-156]. However, the formation of DSBs cannot be induced by $^{\bullet}$OH [29,30,147], which is the expected damaging ROS generated through a Fenton reaction as the result of the reduction of hydrogen peroxide (H_2O_2) by transition metals, the dismutation product of $O_2^{\bullet -}$ [148]. It was proposed that the tentative formation of DSBs associated with the observation of γ-H2AX foci was explained in terms of processing of still unknown oxidatively generated damage during DNA replication [147]. An alternative possibility that however awaits further experiments would implicate the formation of bulky lesions including DNA-protein cross-links and DNA interstrand cross-links [27, 29, 30]. The generation of the latter DNA modifications could be initiated by one-electron oxidation by carbonate radical anion ($CO_3^{\bullet -}$) a reactive species generated by a complex sequence of reactions. This involves recombination of $O_2^{\bullet -}$ and $^{\bullet}$NO giving rise to peroxynitrite that further reacts with CO_2 or bicarbonates with subsequent formation of instable nitrosoperoxycarbonate, the precursor of $CO_3^{\bullet -}$. In that respect it is worth mentioning that one-electron oxidation of nucleobases in the DNA of cells using a UV laser microbeam was found to induce the formation of γ-H2AX foci in non-exposed cells [157]. Furthermore $^{\bullet}$NO has been shown to be implicated in the formation of DSBs in bystander cells upon α-particle irradiation [154].

CONCLUSION

Major progress has been made during the last decade toward a better understanding of the radiation-induced degradation pathways of base and 2-deoxyribose in cellular through the accurate HPLC-ESI-MS/MS measurement of more than 20 oxidation products. It clearly appears that the yield of oxidatively generated base lesions is rather low, being in the range of a few up to 30 lesions per 10^9 bases and per Gy. This is about two or three orders lower than the frequencies of base modifications previously established using the questionable GC-MS and HPLC-MS methods that suffer from major drawbacks. Therefore, it may be stressed that radiation-induced base lesions, if considered as single damage, play at best only a minor biological role since at least 30 Gy are necessary to double the yield of steady-state level of modified bases arising from oxygen metabolism. The predominance of the indirect effects of ionizing radiation mediated by $^{\bullet}OH$ over ionization reactions through direct effects is another key information. It has also been inferred from the characterization of the oxidatively generated damage to cellular DNA that model studies performed in aerated aqueous solutions are highly relevant. It remains to search for the formation in cells of radiation-induced DNA-protein and interstrand cross-links for which mechanistic information involving one-electron oxidation of guanine followed by nucleophilic addition reactions is available. This also applies to the formation of tandem base lesions that are triggered by the initial generation of adjacent peroxyl pyrimidine radicals. Major efforts should also be devoted toward a better understanding of the radiation-induced formation of non-DSB clustered modifications.

REFERENCES

[1] Scholes G, Weiss J, Wheeler CM. Formation of hydroperoxides from nucleic acids by irradiation with x-rays in aqueous systems. *Nature* 1956; 178:157.

[2] Ekert B, Monier R. Structure of thymine hydroperoxides produced by x-irradiation. *Nature* 1959; 184:58-59.

[3] Téoule R, Cadet J. Radiolysis of thymine in aerated aqueous solutions. *J. Chem. Soc. Chem. Commun.* 1971; 1269-1270.

[4] Hahn BS, Wang SY, Flippen JL, Karle IL. Radiation chemistry of nucleic acids. Isolation and characterization of glycols of 1-carbamoylimidazolidone as products of cytosine. *J. Am. Chem. Soc.* 1973; 95, 2711-2712.

[5] Cadet J, Ulrich J, Téoule R. Isomerization and new specific synthesis of thymine glycol. *Tetrahedron* 1975; 31, 2057-2061.

[6] Hariharan PV, Cerutti PA. Repair of gamma-ray-induced thymine damage in Micrococcus radiodurans. 1971; *Nat. New Biol.* 229:247-249.

[7] Hariharan PV, Cerutti PA. Formation and repair of gamma-ray induced thymine damage in Micrococcus radiodurans. *J. Mol. Biol.* 1972; 66:65-81.

[8] Hariharan PV, Cerutti PA (1974) Excision of damaged thymine residues from gamma-irradiated poly(dA-dT) by crude extracts of Escherichia. *Proc. Natl. Acad. Sci. USA* 71:3532-3536.

[9] Remsen JF, Cerutti PA. Deficiency of gamma-ray excision repair in skin fibroblasts from patients with Fanconi's anemia. *Proc. Natl. Acad. Sci. USA* 1976; 73:2419-2423.

[10] Cadet J, Berger M. Radiation-induced decomposition of the purine bases within DNA and related model compounds. *Int. J. Radiat. Biol.* 1985; 47:127-143.

[11] Cadet J, Douki T, Ravanat J-L, Wagner JR. Measurement of oxidatively generated base damage to nucleic acids in cells: facts and artefacts. *Bioanal. Rev.* 2012; 4:55-74.

[12] Floyd RA, Watson JJ, Wong PK, Altmiller DH, Rickard RC. Hydroxyl free radical adduct of deoxyguanosine: sensitive detection and mechanisms of formation. *Free Radic. Res. Commun.* 1986; 1:163-172.

[13] Kasai H, Nishimura S (1984) Hydroxylation of deoxyguanosine at the C-8 position by ascorbic acid and other reducing agents. *Nucleic Acids Res.* 1984; 12:2137-2145.

[14] Kasai H, Crain PF, Kuchino Y, Nishimura S, Ootsuyama A, Tanooka H. Formation of 8-hydroxyguanine moiety in cellular DNA by agents producing oxygen radicals and evidence for its repair. *Carcinogenesis* 1986; 7:1849-1851.

[15] Dizdaroglu M. The use of capillary gas chromatography-mass spectrometry for identification of radiation-induced DNA base damage and DNA base amino acid cross-links. *J. Chromatogr.* 1984; 295:103–121.

[16] Dizdaroglu M. Application of capillary gas chromatography mass spectrometry to chemical characterization of radiation-induced base damage of DNA: implication for assessing DNA repair process. *Anal. Biochem.* 1985; 144:593–603.

[17] Hamberg M, Zhang L-Y. Quantitative determination of 8-hydroxyguanine and guanine by isotope dilution mass spectrometry. *Anal. Biochem.* 1995; 229:336–344.

[18] Ravanat J-L, Turesky RJ, Gremaud E, Trudel LJ, Stadler RH. Determination of 8-oxoguanine in DNA by gas chromatography-mass spectrometry and HPLC-electrochemical detection: overestimation of the background level of the oxidized base by the gas chromatography-mass spectrometry assay. *Chem. Res. Toxicol.* 1995; 8:1039–1045.

[19] Cadet J, Douki T, Ravanat J-L. Artifacts associated with the measurement of oxidized DNA bases. *Environ. Health Perspect.* 1997; 105:1034–1039.

[20] Ravanat J-L, Duretz B, Guiller A, Douki T, Cadet J. Isotope dilution high-performance liquid chromatography–electrospray tandem mass spectrometry assay for the measurement of 8-oxo-7,8-dihydro-2′ -deoxyguanosine in biological samples. *J. Chromatogr. B* 1998; 715:349–356.

[21] Frelon S, Douki T, Ravanat J-L, Pouget J-P, Tornabene C, Cadet J. High-performance liquid chromatography–tandem mass spectrometry measurement of radiation-induced base damage to isolated and cellular DNA. *Chem. Res. Toxicol.* 2000; 13:1002–1010.

[22] Ravanat J-L, Douki T, Duez P, Gremaud E, Herbert K, Hofer T, Lassere L, Saint-Pierre C, Favier A, Cadet J. Cellular background of 8-oxo-7,8-dihydro-2′ -deoxyguanosine: an isotope based method to evaluate artefactual oxidation of DNA during its extraction and subsequent work-up. *Carcinogenesis* 2002; 23:1911–1918.

[23] Chao MR, Yen CC, Hu CW. Prevention of artifactual oxidation in determination of cellular 8-oxo-7,8-dihydro-2′ -deoxyguanosine by isotope-dilution LC-MS/MS with automated solid-phase extraction. *Free Radic. Biol. Med.* 2008; 44:464–473.

[24] Cadet J, Douki T, Ravanat J-L (2011) Measurement of oxidatively generated base damage in cellular DNA. *Mutat. Res.* 2011; 711:3–12.

[25] Wagner JR, Cadet J. Oxidation reactions of cytosine DNA components by hydroxyl radical and one-electron oxidants in aerated aqueous solutions. *Acc. Chem. Res.* 2010; 43:564-571.

[26] Cadet J, Douki T, Ravanat J-L. Oxidatively generated base damage to cellular DNA. *Free Radic. Biol. Med.* 2010; 49:9–21.

[27] Cadet J, Ravanat JL, TavernaPorro M, Menoni H, Angelov D. Oxidatively generated complex DNA damage: tandem and clustered lesions. *Cancer Lett.* 2012; 327:5-15.

[28] Cadet J, Wagner JR. DNA base damage by reactive oxygen species, oxidizing agents, and UV radiation *Cold Spring Harb. Perspect. Biol.* 2013; 5:a012559.

[29] Cadet J, Wagner JR, Shafirovich V, Geacintov NE. One-electron oxidation reactions of purine and pyrimidine bases in cellular DNA. *Int. J. Radiat. Biol.* 2014; 90:423-436.

[30] Cadet J, Wagner JR. Oxidatively generated base damage to cellular DNA by hydroxyl radical and one-electron oxidants: similarities and differences. *Arch. Biochem. Biophys.* 2014 557:47-54.

[31] Ward JF. Biochemistry of DNA lesions. *Radiat. Res. Suppl* 1985: 8:S103-S111.

[32] Ward JF. DNA damage produced by ionizing radiation in mammalian cells: identities, mechanisms of formation, and reparability. *Prog. Nucleic Acid Res. Mol. Biol.* 1988:35:95-125.

[33] Goodhead DT. Initial events in the cellular effects of ionizing radiations: clustered damage in DNA. *Int. J. Radiat. Biol.* 1994; 65:7-17.

[34] O'Neill P, Wardman P. Radiation chemistry comes before radiation biology. *Int. J. Radiat. Biol.* 2009; 85:9-25.

[35] Georgakilas AG, O'Neill P, Stewart RD. Induction and repair of clustered DNA lesions: what do we know so far? *Radiat. Res.* 2013; 180:100-109.

[36] Nikjoo H, O'Neill P, Wilson WE, Goodhead DT. Computational approach for determining the spectrum of DNA damage induced by ionizing radiation. *Radiat. Res.* 2001; 156:577-583.

[37] Stewart RD, Yu VK, Georgakilas AG, Koumenis C, Park JH, Carlson DJ. Effects of radiation quality and oxygen on clustered DNA lesions and cell death. *Radiat. Res.* 2011; 176:587-602.

[38] Nikjoo H, Girard P. A model of the cell nucleus for DNA damage calculations. *Int. J. Radiat. Biol.* 2012; 88:87-97.

[39] Douki T, Martini R, Ravanat JL, Turesky RJ, Cadet J. Measurement of 2,6-diamino-4-hydroxy-5-formamidopyrimidine and 8-oxo-7,8-dihydroguanine in isolated DNA exposed to gamma radiation in aqueous solution. Carcinogenesis 1997; 18:2385-2391.

[40] Berger M, Cadet J. Isolation and characterization of the radiation-induced degradation products of 2'-deoxyguanosine in oxygen-free aqueous solutions. *Z. Naturforsch. B* 1985; 40:1519-1531.

[41] Greenberg MM. The formamidopyrimidines: purine lesions formed in competition with 8-oxopurines from oxidative stress. *Acc. Chem. Res.* 2012; 45:588-597.

[42] Badouard C, Ménézo Y, Panteix G, Ravanat J-L, Douki T, Cadet J, Favier A. Determination of new types of DNA lesions in human sperm. *Zygote* 2008; 16:9–13.

[43] Gedik CM, Wood SG, Collins AR. Measuring oxidative damage to DNA; HPLC and the comet assay compared. *Free Radic. Res.* 1998; 29:609-615.

[44] Pouget JP, Douki T, Richard MJ, Cadet J. DNA damage induced in cells by gamma and UVA radiation as measured by HPLC/GC-MS and HPLC-EC and Comet assay. *Chem. Res. Toxicol.* 2000; 13:541-549.

[45] Sauvaigo S, Petec-Calin C, Caillat S, Odin F, Cadet J. Comet assay coupled to repair enzymes for the detection of oxidative damage to DNA induced by low doses of gamma-radiation: use of YOYO-1, low-background slides, and optimized electrophoresis conditions. *Anal. Biochem.* 2002; 303:107-109.

[46] Pouget JP, Frelon S, Ravanat JL, Testard I, Odin F, Cadet J. Formation of modified DNA bases in cells exposed either to gamma radiation or to high-LET particles. *Radiat. Res.* 2002; 157:589-595.

[47] Douki T, Ravanat J-L, Pouget J-P, Testard I, Cadet J. Minor contribution of direct ionization to DNA base damage induced by heavy ions. *Int. J. Radiat. Biol.* 2006; 82:119-127.

[48] Wagner JR, van Lier JE, Berger M, Cadet J. Synthesis, NMR and decomposition of thymine hydroperoxides: Primary products of radical oxidation of thymidine and DNA. *J. Am. Chem. Soc.* 1994; 116:2235-2242.

[49] Samson-Thibault F, Madugundu GS, Gao S, Cadet J, Wagner JR. Profiling cytosine oxidation in DNA by LC-MS/MS. *Chem. Res. Toxicol.* 2012; 25:1902-1911.

[50] Tremblay S, Douki T, Cadet J, Wagner JR. 2'-Deoxycytidine glycols, a missing link in the free radical-mediated oxidation of DNA. *J. Biol. Chem.* 1999; 274:20833-20838.

[51] Tremblay S, Gantchev T, Tremblay L, Lavigne P, Cadet J, Wagner JR. Oxidation of 2'-deoxycytidine to four interconverting diastereomers of N1-carbamoyl-4,5-dihydroxy-2-oxoimidazolidine nucleosides. *J. Org. Chem.* 2007; 72:3672-3678.

[52] Rivière J, Bergeron F, Tremblay S, Gasparutto D, Cadet J, Wagner JR. Oxidation of 5-hydroxy-2'-deoxyuridine into isodialuric acid, dialuric acid, and hydantoin products. *J. Am. Chem. Soc.* 2004; 126:6548-6549.

[53] Cadet J, Douki T, Ravanat JL. Oxidatively generated damage to the guanine moiety of DNA: mechanistic aspects and formation in cells. *Acc. Chem. Res.* 2008; 41:1075-1083.

[54] Cadet J, Wagner JR. TET enzymatic oxidation of 5-methylcytosine, 5-hydroxymethylcytosine and 5-formylcytosine. *Mutat. Res.* 2014; 764-765:18-35.

[55] Madugundu GS, Cadet J, Wagner JR. Hydroxyl-radical-induced oxidation of 5-methylcytosine in isolated and cellular DNA. *Nucleic Acids Res.* 2014; 42:7450-7460.

[56] Bergeron F, Auvré F, Radicella JP, Ravanat JL. HO* radicals induce an unexpected high proportion of tandem base lesions refractory to repair by DNA glycosylases. *Proc. Natl. Acad. Sci. USA* 2010; 107:5528-5533.

[57] Box HC, Budzinski EE, Dawidzik JB, Wallace JC, Evans MS, Gobey GS. *Radiat. Res.* 1996; 145:641-643.

[58] Bellon S, Ravanat J-L, Gasparutto D, Cadet J (2002) Guanine-thymine intrastrand cross-linked lesion containing oligonucleotides: from chemical synthesis to in vitro enzymatic replication. *Chem. Res. Toxicol.* 15:598-606.

[59] Wang Y. Bulky DNA lesions induced by reactive oxygen species. *Chem. Res. Toxicol* 2008; 21: 276-281.

[60] Xerri B, Morell C, Grand A, Cadet J, Cimino P, Barone V. Radiation-induced formation of DNA intrastrand crosslinks between thymine and adenine bases: A theoretical approach. *Org. Biomol. Chem.* 2006; 4:3986-3992.

[61] V. Labet, C. Morell, A. Grand, J. Cadet, P. Cimino, V. Barone. Formation of cross-linked adducts between guanine and thymine mediated by hydroxyl radical and one-electron oxidation: a theoretical study. *Org. Biomol. Chem.* 2008; 6:3300-3305.

[62] Jiang Y, Hong H, Cao H, Wang Y. In vivo formation and in vitro replication of a guanine-thymine intrastrand cross-link lesion. *Biochemistry* 2007; 46:12757-12763.

[63] Hong H, Wang Y. Derivatization with Girard reagent T combined with LC-MS/MS for the sensitive detection of 5-formyl-2'-deoxyuridine in cellular DNA. *Anal. Chem.* 2007; 79:322-326.

[64] Hong H, Cao H, Wang Y. Formation and genotoxicity of a guanine-cytosine intrastrand cross-link lesion in vivo. *Nucleic Acids Res.* 2007; 35:7118-7127.

[65] Ahnström G, Erixon, K. Measurement of strand breaks by alkaline denaturation and hydroxyapatite chromatography. In Friedberg EC, Hanawalt PC, eds. *DNA Repair. A Laboratory Manual of Research Procedures.* New York: Marcel Dekker, 1981: 403–418.

[66] Collins AR, Oscoz AA, Brunborg G, Gaivão I, Giovannelli L, Kruszewski M, Smith CC, Štětina R. The comet assay: topical issues. *Mutagenesis* 2008; 23:143-151.

[67] von Sonntag C. *Free-Radical-Induced DNA Damage and its Repair. A Chemical Perspective.* Berlin – Heidelberg – New-York: Springer-Verlag, 2006.

[68] Dedon PC. The chemical toxicology of 2-deoxyribose oxidation in DNA. *Chem. Res. Toxicol.* 2008; 21:206-219.

[69] Dizdaroglu M. Clemens von Sonntag and the early history of radiation-induced sugar damage in DNA. *Int. J. Radiat. Biol.* 2014; 90:446-458.

[70] Roginskaya M, Mohseni R, Moore TJ, Bernhard WA, Razskazovskiy Y. Identification of the C4'-oxidized abasic site as the most abundant 2-deoxyribose lesion in radiation-damaged DNA using a novel HPLC-based approach. *Radiat. Res.* 2014; 181:131-137.

[71] Greenberg MM. Abasic and oxidized abasic site reactivity in DNA: enzyme inhibition, cross-linking, and nucleosome catalyzed reactions. *Acc. Chem. Res.* 2014; 47:646-655.

[72] Chen B, Zhou X, Taghizadeh K, Chen J, Stubbe J, Dedon PC. GC/MS methods to quantify the 2-deoxypentos-4-ulose and 3'-phosphoglycolate pathways of 4'-oxidation of 2-deoxyribose in DNA: application to DNA damage produced by gamma radiation and bleomycin. *Chem. Res. Toxicol.* 2007; 20:1701-1708.

[73] . Chan W, Chen B, Wang L, Taghizadeh K, Demott MS, Dedon PC. Quantification of the 2-deoxyribonolactone and nucleoside 5'-aldehyde products of 2-deoxyribose oxidation in DNA and cells by isotope-dilution gas chromatography mass spectrometry: differential effects of gamma radiation and Fe^{2+}-EDTA. *J. Am. Chem. Soc.* 2010; 132:6145-6153.

[74] Chatgilialoglu C, Ferreri C, Terzidis MA. Purine 5',8-cyclonucleoside lesions: chemistry and biology. *Chem. Soc. Rev.* 2011; 40:1368-1382.

[75] Kuraoka I, Bender C, Romieu A, Cadet J, Wood RD, Lindhal T. Removal of oxygen free-radical-induced 5',8-purine cyclodeoxynucleosides from DNA by the nucleotide excision-repair pathway in human cells. *Proc. Natl. Acad. Sci. USA* 2000; 97:3832-3837.

[76] The oxidative DNA lesion 8,5'-(S)-cyclo-2'-deoxyadenosine is repaired by the nucleotide excision repair pathway and blocks gene expression in mammalian cells. Brooks PJ, Wise DS, Berry DA, Kosmoski JV, Smerdon MJ, Somers RL, Mackie H, Spoonde AY, Ackerman EJ, Coleman K, Tarone RE, Robbins JH. The oxidative DNA

lesion 8,5'-(S)-cyclo-2'-deoxyadenosine is repaired by the nucleotide excision repair pathway and blocks gene expression in mammalian cells. *J. Biol. Chem.* 2000; 275: 22355-22362.

[77] Kropachev K, Ding S, Terzidis MA, Masi A, Liu Z, Cai Y, Kolbanovskiy M, Chatgilialoglu C, Broyde S, Geacintov NE, Shafirovich V. Structural basis for the recognition of diastereomeric 5',8-cyclo-2'-deoxypurine lesions by the human nucleotide excision repair system. *Nucleic Acids Res.* 2014; 42:5020-5032.

[78] Boussicault F, Kaloudis P, Caminal C, Mulazzani QG, Chatgilialoglu C. The fate of C5' radicals of purine nucleosides under oxidative conditions. *J. Am. Chem. Soc.* 2008; 130:8377-8385.

[79] Chatgilialoglu C, D'Angelantonio M, Kciuk G, Bobrowski K. New insights into the reaction paths of hydroxyl radicals with 2'-deoxyguanosine. *Chem. Res. Toxicol.* 2011; 24:2200-2206.

[80] Belmadoui N, Boussicault F, Guerra M, Ravanat JL, Chatgilialoglu C, Cadet J. Radiation-induced formation of purine 5',8-cyclonucleosides in isolated and cellular DNA: high stereospecificity and modulating effect of oxygen. *Org. Biomol. Chem.* 2010; 8:3211-3219.

[81] Guerrero CR, Wang J, Wang Y. Induction of 8,5'-Cyclo-2'-deoxyadenosine and 8,5'-cyclo-2'-deoxyguanosine in isolated DNA by Fenton-type reagents. *Chem. Res. Toxicol.* 2013; 26:1361-1366.

[82] D'Errico M, Parlanti E, Teson M, de Jesus BM, Degan P, Calcagnile A, Jaruga P, Bjørås M, Crescenzi M, Pedrini AM, Egly JM, Zambruno G, Stefanini M, Dizdaroglu M, Dogliotti E. New functions of XPC in the protection of human skin cells from oxidative damage. *EMBO J.* 2006; 25:4305-4315.

[83] D'Errico M, Parlanti E, Teson M, Degan P, Lemma T, Calcagnile A, Iavarone I, Jaruga P, Ropolo M, Pedrini AM, Orioli D, Frosina G, Zambruno G, Dizdaroglu M, Stefanini M, Dogliotti E. he role of CSA in the response to oxidative DNA damage in human cells. *Oncogene* 2007; 26:4336-4343.

[84] Regulus P, Spessotto S, Gateau M, Cadet J, Favier A, Ravanat J-L. Detection of new radiation-induced DNA lesions by liquid chromatography coupled to tandem mass spectrometry. *Rapid Commun. Mass Spectrom.* 2004; 18:2223–2228.

[85] Regulus P, Duroux B, Bayle PA, Favier A, Cadet J, Ravanat J-L. Oxidation of the sugar moiety of DNA by ionizing radiation or bleomycin could induce the formation of a cluster DNA lesion. *Proc. Natl. Acad. Sci. USA* 2006; 104:4032–14037.

[86] Povirk LF, Goldberg IH. A role of oxidative DNA sugar damage in mutagenesis by neocarzinostatin and bleomycin. *Biochimie* 1987; 69:815-23.

[87] Sczepanski JT, Jacobs AC, Greenberg MM. Self-promoted DNA interstrand cross-link formation by an abasic site. *J. Am. Chem. Soc.* 2008; 130:9646–9647.

[88] Sczepanski JT, Jacobs AC, Majumdar A, Greenberg MM. Scope and mechanism of interstrand cross-link formation by the C40-oxidized abasic. *J. Am. Chem. Soc.* 2009; 131: 11132–11139.

[89] Douki T, Angelov D, Cadet J. UV laser photolysis of DNA: effect of duplex stability on charge-transfer efficiency. *J. Am. Chem. Soc.* 2001; 123:11360-11366.

[90] Douki T, Ravanat J-L, Angelov D, Wagner JR, Cadet J. Effects of duplex stability on charge transfer efficiency within DNA. *Top. Curr. Chem.* 2004; 236:1–25.

[91] Kasai H, Yamaizumi Z, Berger N, Cadet J. Photosensitized formation of 7,8-dihydro-8-oxo-2'-deoxyguanosine (8-hydroxy-2'-deoxyguanosine) in DNA by riboflavin. A non singlet oxygen mediated reaction. *J. Am. Chem. Soc.* 1992; 114:9692-9694.

[92] Rokhlenko Y, Cadet J, Geacintov NE, Shafirovich V. Mechanistic aspects of hydration of guanine radical cations in DNA. *J. Am. Chem. Soc.* 2014; 136:5956-5962.

[93] Schuster GB. Long-range charge transfer in DNA: transient structural distortions control the distance dependence. *Acc. Chem. Res.* 2000; 33:253-260.

[94] Kawai K, Majima T. G-C content independent long-range charge transfer through DNA. *Top. Curr. Chem.* 2011; 298:129-142.

[95] Madugundu GS, Wagner JR, Cadet J, Kropachev K, Yun BH, Geacintov NE, Shafirovich V. Generation of guanine-thymine cross-links in human cells by one-electron oxidation mechanisms. *Chem. Res. Toxicol.* 2013; 26:1031–1033.

[96] Yun BH, Geacintov NE, Shafirovich V. Generation of guanine-thymidine cross-links in DNA by peroxynitrite/carbon dioxide. *Chem. Res. Toxicol.* 2011; 24:1144 –1152.

[97] Ding S, Kropachev K, Cai Y, Kolbanovskiy M, Durandina SA, Liu Z, Shafirovich V, Broyde S, Geacintov NE. Structural, energetic and dynamic properties of guanine(C8)-thymine(N3) cross-links in DNA provide insights on susceptibility to nucleotide excision repair. *Nucleic Acids Res.* 2012; 40:2506-2517.

[98] Dewey WC, Wong RS, Albright N. Pulsed-field gel electrophoretic migration of DNA broken by X irradiation during DNA synthesis: experimental results compared with Monte Carlo calculations. *Radiat. Res.* 1997; 148:413-420.

[99] Olive PL, Banáth JP. The comet assay: a method to measure DNA damage in individual cells. *Nat. Protocol.* 2006; 1:23–29.

[100] Stap J, Krawczyk PM, Van Oven CH, Barendsen CM, Essers J, Kanaar R, Aten JA. Induction of linear tracks of DNA double-strand breaks by α-particle irradiation of cells. *Nat. Methods* 2008; 5:261-266.

[101] Leatherbarrow EL, Harper JV, Cucinotta FA, O'Neill P. Induction and quantification of gamma-H2AX foci following low and high LET-irradiation. *Int. J. Radiat. Biol.* 2006; 82:111-118.

[102] Bonner WM, Redon CE, Dickey JS Nakamura AJ, Sedelnikova OA Solier S, Pommier Y. Gamma-H2AX and cancer. *Nat. Rev. Cancer* 2008; 8:957–967.

[103] Ivashkevich AN, Martin OA, Smith AJ, Redon CE, Bonner WM, Lobachevsky PN. γ-H2AX as a measure of DNA damage: a computational approach to automatic analysis. *Mutat. Res.* 2011; 711:49–60.

[104] Cleaver JE. γ-H2AX biomarker of damage or functional participant in DNA repair "all that glitters is not gold". *Photochem. Photobiol.* 2011; 87:1230–1239.

[105] Semenenko VA, Stewart RD, Fast Monte Carlo simulation of DNA damage formed by electron and light ions. *Phys. Med. Biol.* 2006; 51:1693-1706.

[106] Alloni D, Campa A, Friedland W, Mariotti L, Ottolenghi A. Track structure, radiation quality and initial radiobiological events: considerations based on the PARTRAC code experience. *Int. J. Radiat. Biol.* 2012; 88:77-86.

[107] Alloni D, Campa A, Friedland W, Mariotti L, Ottolenghi A. Integration of Monte Carlo simulations with PFGE experimental data yields constant RBE of 2.3 for DNA double-strand break induction by nitrogen ions between 125 and 225 keV/μm LET. *Radiat. Res.* 2013; 179:690-697.

[108] Desai N, Davis E, O'Neill P, Durante M, Cucinotta FA, Wu,H. Immunofluorescence detection of clustered γ-H2AX foci induced by HZE-particle radiation. *Radiat. Res.* 2005; 164:518-522.
[109] Chiruvella KK, Liang Z, Wilson TE. Repair of double-strand breaks by end joining. *Cold Spring Harb. Perspect. Biol.* 2013; 5:a012757.
[110] Davis AJ, Chen DJ. DNA double strand break repair via non-homologous end-joining. *Transl. Cancer Res.* 2013; 2:130-143.
[111] Kakarougkas A, Jeggo PA. DNA DSB repair pathway choice: an orchestrated handover mechanism. *Br. J. Radiol.* 2014; 87:20130685.
[112] Asaithamby A, Uematsu N, Chatterjee A, Story, MD, Burma S, Chen DJ. Repair of HZE-particle-induced double-strand breaks in normal human fibroblasts. *Radiat. Res.* 2008; 169:437-446.
[113] Li Y, Reynolds P, O'Neill P, Cucinotta FA. Modeling damage complexity-dependent non-homologous end-joining repair pathway. *PLoS One* 2014; 9:e85816.
[114] Stewart RD, Yu VK, Georgakilas AG, Koumenis C, Park JH, Carlson DJ. Effects of radiation quality and oxygen on clustered DNA lesions and cell death. *Radiat. Res.* 2011; 176:587-602.
[115] Georgakilas AG, Bennett PV, Wilson III DM, Sutherland BM. Processing of bistranded abasic DNA clusters in γ-irradiated human hematopoietic cells. *Nucleic Acids Res.* 2004; 32:5609-5620.
[116] Hada M, Georgakilas AG. Formation of clustered DNA damage after high-LET irradiation: a review. *J. Radiat. Res.* 2008; 49:203-210.
[117] Georgakilas A. Detection of clustered DNA lesions: Biological and clinical applications. *World J. Biol. Chem.* 2011; 2:173-176.
[118] Tsao D, Kalogerinis P, Tabrizi I, Dingfelder M, Stewart, RD, Georgakilas AG. Induction and processing of oxidative clustered DNA lesions in ^{56}Fe-ion-irradiated human monocytes. *Radiat. Res.* 2007; 168:87-97.
[119] Francisco DC, Peddi P, Hair JM, Flood BA, Cecil AM, Kalogerinis PT, Sigounas G, Georgakilis AG. Induction and processing of complex DNA damage in human breast cancer cells MCF-7 and non-malignant MCF-10A cells. *Free Radic. Biol. Med.* 2008; 44, 558-569.
[120] Boucher D, Testard I, Averbeck D. Low levels of clustered oxidative DNA damage induced at low and high LET irradiation in mammalian cells. *Radiat. Environ. Biophys.* 2006; 45:267-276.
[121] Pimblott SM, LaVerne JA. Production of low-energy electrons by ionizing radiation. *Radiat. Phys. Chem.* 2007; 76:1244-1247.
[122] Sanche L. Low energy electron-driven damage in biomolecules. *Eur. Phys. J. D* 2005; 35:367-390.
[123] Zheng Y, Cloutier P, Hunting DJ, Wagner JR, Sanche L. Glycosidic bond cleavage of thymidine by low-energy electrons. *J. Am. Chem. Soc.* 2004; 126:1002-1003.
[124] Zheng Y, Cloutier P, Hunting DJ, Sanche L, Wagner JR. Chemical basis of DNA sugar-phosphate cleavage by low-energy electrons. *J. Am. Chem. Soc.* 2005; 127:16592-16598.
[125] Zheng Y, Wagner JR, Sanche L. DNA damage induced by low-energy electrons: Electron transfer and diffraction. *Phys. Rev. Lett.* 2006; 96:208101.

[126] Zheng Y, Cloutier P, Hunting DJ, Wagner JR, Sanche L. Phosphodiester and N-glycosidic bond cleavage in DNA induced by 4-15 eV electrons - art. no. 064710. *J. Chem. Phys.* 2006; 124:64710.

[127] Li ZJ, Zheng Y, Cloutier P, Sanche L, Wagner JR. Low energy electron induced DNA damage: Effects of terminal phosphate and base moieties on the distribution of damage. *J. Am. Chem. Soc.* 2008; 130:5612-5613.

[128] Li ZJ, Cloutier P, Sanche L, Wagner JR. Low-energy electron-enduced DNA damage: effect of base sequence in oligonucleotide trimers. *J. Am. Chem. Soc.* 2010; 132, 5422-5427.

[129] Park Y, Li ZJ, Cloutier P, Sanche L, Wagner JR. DNA Damage induced by low-energy electrons: conversion of thymine to 5,6-dihydrothymine in the oligonucleotide trimer TpTpT. *Radiat. Res.* 2011; 175:240-246.

[130] Li Z, Cloutier P, Sanche L, Wagner JR. Low-energy electron-induced damage in a trinucleotide containing 5-bromouracil. *J. Phys. Chem. B* 2011; 115:13668-13673.

[131] Alizadeh E, Cloutier P, Hunting D, Sanche L. Soft X-ray and low energy electron-induced damage to DNA under N_2 and O_2 atmospheres. *J. Phys. Chem. B* 2011; 115:4523-4531.

[132] Madugundu G S, Park Y, Sanche L, Wagner JR. Radiation-induced formation of 2',3'-dideoxyribonucleosides in DNA: A potential signature of low-energy electrons. *J. Am. Chem. Soc.* 2012; 134:17366-17368.

[133] Park Y, Polska K, Rak J, Wagner JR, Sanche L. Fundamental mechanisms of DNA radiosensitization: Damage induced by low-energy electrons in brominated oligonucleotide trimers. *J. Phys. Chem. B* 2012; 116:9676-9682.

[134] Alizadeh E, Sanche L. Precursors of solvated electrons in radiobiological physics and chemistry. *Chem. Rev.* 2012; 112:5578-5602.

[135] Park Y, Peoples AR, Madugundu GS, Sanche L, Wagner JR. Side-by-side comparison of DNA damage induced by low-energy electrons and high-energy photons with solid TpTpT trinucleotide. *J. Phys. Chem. B* 2013; 117:10122-10131.

[136] Alizadeh E, Sanz AG, Madugundu GS, García G, Wagner JR, Sanche L. Thymidine decomposition induced by low-energy electrons and soft X-rays under N_2 and O_2 atmospheres. *Radiat. Res.* 2014; 181:629-640.

[137] Martin, F., Burrow, P. D., Cai, Z. L., Cloutier, P., Hunting, D., and Sanche, L. DNA strand breaks induced by 0-4 eV electrons: The role of shape resonances, *Phys. Rev. Lett.* 2004; 93:068101.

[138] Nagasawa H, Little JB. Induction of sister chromatid exchanges by extremely low doses of alpha-particles. *Cancer Res.* 1992; 52:6394-6396.

[139] Kadhim MA, Macdonald DA, Goodhead DT, Lorimore SA, Marsden SJ, Wright EG. Transmission of chromosomal instability after plutonium alpha-particle irradiation. *Nature* 1992; 355:738-740.

[140] Morgan WF, Hartmann A, Limoli CL, Nagar S, Ponnaiya B. Bystander effects in radiation-induced genomic instability. *Mutat. Res.* 2002; 504:91-100.

[141] Morgan WF, Sowa MB. Non-targeted bystander effects induced by ionizing radiation. *Mutat. Res.* 2007; 616:159-164.

[142] Prise KM, O'Sullivan JM. Radiation-induced bystander signalling in cancer therapy. *Nat. Rev. Cancer* 2009; 9:351-360.

[143] Sjostedt S, Bezak E. Non-targeted effects of ionising radiation and radiotherapy. *Australas. Phys. Eng. Sci. Med.* 2010; 33:219-231.

[144] Hei TK, Zhou H, Chai Y, Ponnaiya B, Ivanov VN. Radiation induced non-targeted response: mechanism and potential clinical implications. *Curr. Mol. Pharmacol.* 2011; 4:96-105.

[145] Sprung CN, Ivashkevich A, Forrester HB, Redon CE, Georgakilas A, Martin OA. Oxidative DNA damage caused by inflammation may link to stress-induced non-targeted effects. *Cancer Lett.* 2013. doi: 10.1016/j.canlet.2013.09.008.

[146] Wang H, Yu KN, Hou J, Liu Q, Han W. Radiation-induced bystander effect: Early process and rapid assessment. *Cancer Lett.* 2013; doi:10.1016/j.canlet.2013.09.031.

[147] Klammer H, Mladenov E, Li F, Iliakis G. Bystander effects as manifestation of intercellular communication of DNA damage and of the cellular oxidative status. *Cancer Lett.* 2013; doi: 10.1016/j.canlet.2013.12.017.

[148] Hei TK, Zhou H, Ivanov VN, Hong M, Lieberman HB, Brenner DJ, Amundson SA, Geard CR. Mechanism of radiation-induced bystander effects: a unifying model. *J. Pharm. Pharmacol.* 2008; 60:943–950.

[149] Cadet J, Loft S, Olinski R, Evans MD, Bialkowski K, Richard Wagner J, Dedon PC, Møller P, Greenberg MM, Cooke MS. Biologically relevant oxidants and terminology, classification and nomenclature of oxidatively generated damage to nucleobases and 2-deoxyribose in nucleic acids. *Free Radic. Res.* 2012; 46:367-381.

[150] Spitz DR, Azzam EI, Li JJ, Gius D. Metabolic oxidation/reduction reactions and cellular responses to ionizing radiation: a unifying concept in stress response biology. *Cancer Metast. Rev.* 2004; 23:311–322.

[151] Morgan WF, Sowa MB. Non-targeted effects induced by ionizing radiation: Mechanisms and potential impact on radiation induced health effects. *Cancer Lett.* 2013; doi: 10.1016/j.canlet.2013.09.009.

[152] Georgakilas AG. Bystander and non-targeted effects: A unifying model from ionizing radiation to cancer. *Cancer Lett.* 2014; doi: 10.1016/j.canlet.2014.03.032.

[153] Yang H, Asaad N, Held KD. Medium-mediated intercellular communication is involved in bystander responses of X-ray-irradiated normal human fibroblasts. *Oncogene* 2005; 24:2096-2103.

[154] Han W, Chen S, Yu KN, Wu L. Nitric oxide mediated DNA double strand breaks induced in proliferating bystander cells after alpha-particle irradiation. *Mutat. Res.* 2010; 684:81-89.

[155] Klammer H, Kadhim M, Iliakis G. Evidence of an adaptive response targeting DNA nonhomologous end joining and its transmission to bystander cells. *Cancer Res.* 2010; 70:8498-8506.

[156] Ojima M, Furutani A, Ban N, Kai M. Persistence of DNA double-strand breaks in normal human cells induced by radiation-induced bystander effect. *Radiat. Res.* 2011; 175:90-96.

[157] Botchway SW, Reynolds P, Parker AW, O'Neill P. Laser-induced radiation microbeam technology and simultaneous real-time fluorescence imaging in live cells. *Methods Enzymol.* 2012; 504:3-28.

In: Biological Responses, Monitoring and Protection … ISBN: 978-1-63463-852-4
Editor: K. P. Mishra

Chapter 2

STEM CELLS AND REGENERATION OF IRRADIATED TISSUES

Mohi Rezvani*[*], *PhD
Reader Emeritus, Faculty of Clinical Medicine,
University of Oxford, UK

ABSTRACT

The recent advances in the field of stem cell research and regenerative medicine introduce a new and very promising modality in the treatment of degenerative diseases. The findings open a new field of research for the treatment of radiation-induced normal tissue lesions. Normal tissue toxicity is a common side effect of radiotherapy treatment of cancer patients. It is also a major concern in populations subjected to accidental exposure to ionising radiation. At present, there is not a standard protocol for the treatment of radiation-induced normal tissue lesions. Stem cell transplantation or administration of microparticles extracted from stem cells appears to be a promising modality for the treatment of radiation lesions.

1. INTRODUCTION

Stem cell transplantation stems from works carried out in radiobiology with haematopoietic cell transplantation. Jacobson et al. [1] reported that shielding the spleen of a total body irradiated mouse increased the probability of its survival. This was followed by Lorenz et al. [2] who reported modification of radiation injury in mice and guinea pigs by bone marrow infusion. Following these observations, stem cells have been used in medical practice for decades for the treatment of both malignant and non-malignant diseases such as leukaemia, lymphoma and certain types of anaemia in allogenic bone marrow transplantation procedures [3]. After myeloabalation, induced by radiation or chemotherapy, whole allogenic marrow or stem cells extracted from marrow are transplanted to reconstruct the

[*] mohi@me.com.

haematopoietic tissues of the host. A historical review of the development of bone marrow transplantation is given by Thomas [4].

During the initial period of research cell transplantation studies were mainly confined to bone marrow transplantation and colony forming ability of the marrow stromal cells [5]. These cells were also named plastic-adherent cells, colony forming-unit fibroblasts or mesenchymal stem cells that were described as cells responsible for transferring the microenvironment of the haematopoietic tissues and were capable of differentiating into a number of tissues. Isolation of stem cells from adipose tissue, dental pulp, lamina properia, peripheral blood, urine and other tissues followed.

2. Types of Stem Cells

Stem cell is undifferentiated cell that can divide through mitosis, self-renew itself to produce more stem cells and differentiate into specialized cells of the bodily organs. There are three main sources of stem cells that are being currently studied for the treatment of human ailments including radiation damage. These are embryonic stem cells that are located in the inner cell mass of blastocysts, adult stem cells that are practically available in all tissues and induced pluripotent stem cells. Mesenchymal stem cells (MSC), a type of adult stem cell, are the most extensively studied adult stem cells. These are mainly derived from bone marrow (BM-MSC) or adipose tissue (ADSC) but they reside virtually in all postnatal tissues [6]. MSCs should have multipotent differentiation potency to differentiate into osteoblasts, chondrocytes, and adipocytes. MSCs were also shown to differentiate into the ectodermal lineage such as neurons [7] and endodermal lineage such as myocytes [8, 9], cardiomyocytes [10, 11] and hepatocytes [12]. It has also been shown that MSCs do not give rise to tumours [7] and demonstrate antitumour effect [13].

Homing of engrafted MSCs are directed by the cues from the host. This appears to be regulated through a series of signals from damaged tissue and under the influence of local microenvironment. Transplanted MSCs respond to injury and secrete diverse trophic and immunomodulatory factors that bring about subsequent tissue repair and regeneration. Regardless of the cause of damage, almost all studies demonstrate that transplanted MSCs migrate specifically to the damaged tissues. MSCs transplanted intravenously to rats with myocardial damage homed to the infarct region of the heart but in uninjured rats the transplanted cells homed to the bone marrow [10]. Systemically transplanted MSCs to total body irradiated mice, migrated to liver, bone, bone marrow, gut, hearth, kidney, brain, lung and stomach. They were found up to 3 months post irradiation in gut, lung, bone marrow, and heart. Transplanted cells were scarcely detectable in other body areas and in the blood. After abdominal radiation exposure these cells engrafted preferentially in the abdomen: liver, spleen, stomach, kidney and gut. [14, 15]. These observations support the hypothesis that MSC can home to sites of radiation injury.

MSCs secrete many growth factors and proteases [16] including vascular endothelial growth factor (VEGF), hepatocyte growth factor (HGF), insulin-like growth factors (IGF)-1, stromal cell-derived factor (SDF), basic fibroblast growth factor (bFGF), matrix metalloproteinases (MMP), transforming growth factor (TGF)-β, and platelet derived growth factor (PDGF).

Among adult stem cells BM-MSCs, that were the first type of stem cells applied in regenerative medicine, appear to be the most used in the treatment of radiation lesions; mainly in allogenic settings. However, aspirating bone marrow from the patient is an invasive procedure. Alternatively, adipose tissue appears as an attractive source of MSCs for stem cell therapy. Expansion rate, differentiation and immunomodulatory potencials of BM-MSCs and adipose tissue derived mesenchymal stem cells (ADSCs) are comparable [7]. The frequency of stem cells in adipose tissue range from 1in 100 to 1 in 1500 cells that is much higher than the frequency of MSCs in bone marrow [17]. Around 100,000 MSCs can be obtained from each gram of fat and it has been shown that ADSCs are less affected by age than BM-MSCs [18].

3. Stem Cell Treatment

There are a number of studies that report the treatment of radiation-induced normal tissue lesions by autologous fat grafting. Rigotti et al. [19] reported the results of twenty patients suffering from severe side effects of radiotherapy. These patients were treated with autologous fat grafting that resulted in a systematic improvement or remission of symptoms in all evaluated patients. Phulpin et al. [20] treated aesthetic defects caused by radiotherapy in head and neck cancer patients by fat grafting. The treatment resulted in improved quality of life and amelioration of radiation-induced lesions. Kim et al. [21] treated six patients by serial autologous fat grafting after irradiation to the orbit and enucleation. Four of the six patients were successfully fitted for orbital prostheses and the authors concluded that autologous fat grafting could be used for improved ocular implantation to the irradiated enucleated orbit. Serra-Renom et al. [22] and Salgarello et al. [23] also reported a favourable outcome and formation of new subcutaneous tissue after post mastectomy fat grafting in breast-irradiated patients. Breast irradiated patients do not respond favourably to allopathic breast reconstruction and the procedure is usually associated with complications [24]. The favourable results obtained by fat grafting can be attributed to MSC component of the transplanted fat. Fat contains much more MSCs than bone marrow.

Bertho et al. [25] studied the effects of bone marrow derived mononuclear cells in the treatment of radiation-induced pancytopenia in a non-human primate model. The animals were irradiated with 8 Gy total body dose and the cells were transplanted 24 hours after irradiation. Results showed that stem cell transplantation shortened the period and the severity of pancytopenia and improved hematopoietic recovery. In another report the same authors [26] reported a case of a 27 year old Chilean radiation accident victim who received a major gamma dose to his buttock. This patient was treated by excision of the irradiated buttock and wound closure by a skin allograft that did not last very long and got infected immediately. Later at day 90 after irradiation the patient received a skin autograft together with 168 million autologus BM-MSC followed by 226 million BM-MSCs nine days later. A complete healing, 75 days after stem cell transplantation, was reported.

Among non-haematopoietic tissues, central nervous system has been the first to be treated with stem cells. Groves et al. [27] and Franklin et al. [28] demonstrated that transplantation of oligodendrocyte progenitor cells resulted in extensive re-myelination of areas of persistent demyelination in the adult rat's spinal cord.

Later a 12-mm section of the cervical spinal cord of 5-week-old rats was irradiated with a single dose of 22 Gy of Co-60 γ- rays [29]. This dose is known to produce myelopathy in rats within 6 months of irradiation. These authors tested the effect of two different types of adult stem cells. The cells were transplanted intradurally, 90 days after spinal cord irradiation and it was shown that the paralysis-free survival rates of rats that received stem cell transplantation were a third higher than historical controls at 183 days after irradiation. It was concluded that transplantation of neural stem cells 90 days after irradiation significantly ameliorated the expression of radiation-induced myelopathy in rats. Efficacy of stem cell transplantation on radiation induced CNS lesions was supported by Acharya et al. [30, 31] who demonstrated that engrafting human neural stem cells to cranial irradiated athymic nude rats significantly ameliorated cognitive decline compared with sham irradiated animals.

Mouiseddine [32] studied the effects of stem cell transplantation on radiation induced liver damage. In this study NOD–SCID mice were subjected to a 10.5 Gy abdominal dose and the effects of human bone marrow derived mesenchymal cells (hBM-MSC) on biological and histological markers of liver injury were assessed. It was shown that irradiation induced significant apoptosis in the endothelial layer of vessels and elevated the transaminases ALT (353.9 ± 37.5 vs. 108.6 ± 18.9 U/I) and AST (40.3 ± 4.9 vs. 20.3 ± 0.8 U/I) levels after 5 days ($p < 0.001$). However, transplantation of hBM-MSCs by infusion almost completely removed the apoptosis from 29± 7 in radiation alone group to 3±2 in stem cell transplanted animals. Transaminases values at 5 days after irradiation were also significantly ($p < 0.05$) decreased in stem cell transplanted animals; AST (198.8 ± 43.1 U/I) and ALT (14.9 ± 4.7 U/I).

Xue et al. [33] demonstrated that both MSCs and gene-modified MSCs migrated into injured lung in lung irradiated mice. MSC transplantation ameliorated lung injury assessed by survival, histopathology, malondialdehyde (MDA) assay, hydroxyproline, plasma cytokines, expression of connective tissue growth factor (CTGF) and α-smooth muscle actin (a-SMA). The authors observed that the number of transplanted MSCs in the lungs of host animals were too low to explain the observed level of improvements and attributed the beneficial effects of MSCs to their paracrine effect. This was supported by in vitro studies where they demonstrated that MSC conditioned medium attenuated the inflammatory response, radiation-induced apoptosis and DNA damage in irradiated cells.

Semont et al. [34] transplanted hBM-MSC into gut irradiated NOD/SCID mice by infusion. In this study mice were subjected to a total body irradiation (TBI) at a dose of 3.5 Gy immediately followed by the delivery of local irradiation to the abdomen at a dose of 5 Gy resulting a total dose of 8.5 Gy to the abdominal region. Animals were killed at 3-120 days after irradiation and small intestine was collected for analysis. It was demonstrated that hBM-MSC transplantation improved the recovery of the small intestine, both structurally and functionally and extended the life of mice. This was attributed to the improvements in the renewal capability of small intestinal epithelium. hBM-MSC infusion also assisted in re-establishment of cellular homeostasis through increasing endogenous proliferation processes and inhibition of radiation-induced apoptosis of small intestinal epithelial cells.

Linard et al. [35] studied the effects of autologous BM-MSCs in the management of radiation-induced proctitis in a pig model. Irradiated pigs received 2×10^6 MSCs per kilogram body weight intravenously on days 27, 34 and 41 after irradiation. Repeated administrations of MSCs ameliorated the development of radiation-induced fibrosis. While stem cell treatment reduced expression of inflammation and IL-10 it had a beneficiary effect on controlling systemic inflammation. The systemic effect is more likely to be due to the paracrine effect of stem cells.

Agay et al. [36] reported some beneficial effects of stem cell transplantation in a minipig model. These authors irradiated the skin of minipigs with 50 and 60 Gy Co-60 gamma rays and followed up for over 30 weeks. These authors demonstrated that intradermal injection of autologus BM-MSCs, 2-3 times a week from 4-14 weeks after irradiation, led to improved vascularisation. However, hBM-MSCs, intravenously transplanted into NOD/SCID mouse, showed significant improvement in the development and healing of radiation-induced skin damage [37]. The hind leg skin of mice was irradiated with 30 Gy single dose of Co-60 gamma rays and stem cell transplantation was carried out 24 hours after irradiation. It was demonstrated that the skin of mice transplanted with stem cells was partially healed at 6 weeks. This was followed by a complete healing of the epidermis at 8 weeks after irradiation. Only a partial healing of skin lesion was observed in radiation only group of animals at 8 weeks. Recently, Huang et al. [38] showed that ADSC transplantation increased angiogenesis and granulation and accelerated the healing of irradiated wounds in rats.

In an attempt to study the effect of endogenous stem cells, Lombaert et al. [39] assessed the effects of autologous BM-MSC mobilized by administration of granulocyte stimulating factor (G-CSF) in regeneration of irradiated salivary glands and reported a significant improvement in the gland weight, number of ancinar cells and salivary flow rate.

4. Discussion

The target cell theory of radiation injury, that was developed on the basis of earlier studies by Puck and Marcus [40], considers radiation-induced cell loss as the cardinal cause of normal tissue damage or tumour ablation by radiation. In fact, cell loss and regeneration is a natural process that happens in almost all tissues. Cells lost due to ageing or other insults are continually replaced by proliferation of stem cells and differentiation of tissue specific progenitor cells. The hemostasis state of normal tissues is a result of the balance between cell loss and cell production. However, after irradiation, while the cell loss continues the stem or progenitor cells killed or damaged by irradiation fail to produce replacements for the lost cells. The number of cells lost exceeds the number of cells produced and the balance tips over the cell loss. This results in a deficit in the number of differentiated cells. When the cell number reduces below a critical level, a lesion or a functional insufficiency manifests. The evidence reviewed above and those quoted by the others [41, 42, 43] effectively support the idea that radiation damage can be ameliorated by transplantation of stem cells. However, majority of the papers conclude that transplanted cells do not contribute significantly by direct differentiation and replacement of the damaged tissues. Most authors either could not trace the transplanted cells in the damaged tissues or reported a very low level of engraftment that could not justify the functional improvement observed after stem cell transplantation. Transplantation of stem cells 90 days after induction of radiation-induced myelopathy reduced the paralysis by third but no transplanted cells were found in the spinal cord [29]. Mouiseddine et al. [32] observed improvements in liver irradiated animals after stem cell transplant but did not find any MSCs in the liver. Xue et al. [33] demonstrated that the number of transplanted MSCs in the lungs of host animals were too low to explain the observed level of improvements. These observations were attributed to paracrine effect. This was supported by their in vitro studies where they demonstrated that MSC conditioned medium attenuated the inflammatory response, radiation-induced apoptosis and DNA damage in irradiated cells.

All these findings point to the fact that transplanted stem cells exert their action in a paracrine fashion. Improvements in tissue response to irradiation after stem cell transplantation can mainly be due to the release of trophic factors by transplanted cells. Tissue repair is promoted by production and release of soluble factors that may inhibit inflammation and stimulate the endogenous stem cells to repair the damaged tissues. Reduction in radiation-induced complication and improvement in formation of subcutaneous tissue after post mastectomy fat grafting in breast irradiated patients [22, 23] can well be attributed to the stem cell component of the fat graft that acted also in a paracrine fashion. Mouiseddine et al. [32] suggested that cytokines and growth factors produced by the transplanted MSCs stimulate the regeneration of damaged organs by an indirect effect. MSCs secret angiogenic and antiapoptotic factors [44] and express numerous genes of neurotrophins and extracellular matrix proteins required for the nerve growth and myelination. Induction of neural differentiation of ADSCs enhances production of brain-derived neurotrophic factor (BDNF) as well as ability of these cells to induce nerve fiber growth via BDNF production [45]. Bioactive factors secreted by MSCs are both immunomodulatory and trophic. In damaged tissues the trophic component of secretions inhibits ischaemia-induced apoptosis and scarring, stimulates angiogenesis and the mitosis of tissue intrinsic progenitor cells. The immunomodulation component inhibits lymphocyte surveillance of the injured tissue [42]. MSCs also release bioactive factors such as VEGF, IGF-1, EGF, keratinocyte growth factor, angiopoietin-1, stromal derived factor-1, macrophage inflammatory protein-1α and β and erythropoietin [46]. Chiellini et al. [47] identified a total of 73 proteins released in the secretome of human ADSCs. This included proteases, protease inhibitors, extracellular matrix components, cytosklethal components, anti-inflammatory/antioxidant proteins, heat shock proteins and metabolic enzymes.

There are number of excellent reviews on paracrine action of stem cells and their role in tissue regeneration [16, 48, 49]. Paracrine effect of stem cells is exerted through the intercellular communication by the exchange of protein and RNA-containing microparticles and exosomes. Microparticles are a hetrogeous population of small vesicles with a diameter of up to1000 nanometer. These are membrane-coated particles released during the activation of cells or apoptosis [50, 51]. Microparticles of 100-1000 nm are named microvesicles. Exosomes are smaller microparticles (40–100 nm). Microparticles have a bilipid membrane with the same orientation as plasma membrane and a cargo that includes both proteins and RNA. MSCs are known for secreting therapeutic paracrine factors and RNA-containing microparticles. Chen et al. [52] demonstrated that MSC conditioned medium contained small RNAs encapsulated in cholesterol-rich phospholipid vesicles and concluded that MSCs could facilitate miRNA-mediated intercellular communication by secreting microparticles. Micropartiles released from MSCs have protective effect against ischaemia-reperfusion-induced acute and chronic kidney injury [53]. Janowska-Wieczorek et al. [54] demonstrated that mouse bone marrow cells covered with platelet derived microparticles engrafted lethally irradiated mice significantly faster than those not covered, indicating that platelet derived microparticles play an important role in the homing of the transplanted cells.

The potentials of mciroparticles secreted from activated MSCs and cell free approach should be further investigated in the treatment of radiation-induced injuries.

References

[1] Jacobson LO, Marks EK, Robson MJ, Gaston EO, Zirkle RE. Effect of spleen protection on mortality following x-irradiation. *Journal of Laboratory and Clinical Medicine*. 1949;34:1538 – 43.

[2] Lorenz E, Uphoff D, Reid TR, Shelton E. Modification of irradiation injury in mice and guinea pigs by bone marrow injections. *J Natl Cancer Inst*. 1951;12(1):197-201.

[3] Friedrichs B, Tichelli A, Bacigalupo A, Russell NH, Ruutu T, Shapira MY, et al. Long-term outcome and late effects in patients transplanted with mobilised blood or bone marrow: a randomised trial. *Lancet Oncol*. 2010;11(4):331-8.

[4] Thomas ED. A history of haemopoietic cell transplantation. *Br J Haematol*. 1999;105(2):330-9.

[5] Friedenstein AJ, Chailakhyan RK, Latsinik NV, Panasyuk AF, Keiliss-Borok IV. Stromal cells responsible for transferring the microenvironment of the hemopoietic tissues. Cloning in vitro and retransplantation in vivo. *Transplantation*. 1974;17(4):331-40.

[6] Hennrick KT, Keeton AG, Nanua S, Kijek TG, Goldsmith AM, Sajjan US, et al. Lung cells from neonates show a mesenchymal stem cell phenotype. *American journal of respiratory and critical care medicine. 2007*;175(11):1158-64.

[7] Ra JC, Shin IS, Kim SH, Kang SK, Kang BC, Lee HY, et al. Safety of intravenous infusion of human adipose tissue-derived mesenchymal stem cells in animals and humans. *Stem Cells Dev*. 2011;20(8):1297-308.

[8] Gerhart J, Bast B, Neely C, Iem S, Amegbe P, Niewenhuis R, et al. MyoD-positive myoblasts are present in mature fetal organs lacking skeletal muscle. *J Cell Biol*. 2001;155(3):381-92.

[9] Zuk PA, Zhu M, Mizuno H, Huang J, Futrell JW, Katz AJ, et al. Multilineage cells from human adipose tissue: implications for cell-based therapies. *Tissue Eng*. 2001;7(2):211-28.

[10] Saito T, Kuang JQ, Bittira B, Al-Khaldi A, Chiu RC. Xenotransplant cardiac chimera: immune tolerance of adult stem cells. *The Annals of thoracic surgery*. 2002;74(1):19-24; discussion.

[11] Wang HS, Hung SC, Peng ST, Huang CC, Wei HM, Guo YJ, et al. Mesenchymal stem cells in the Wharton's jelly of the human umbilical cord. *Stem Cells*. 2004;22(7): 1330-7.

[12] Lange C, Bassler P, Lioznov MV, Bruns H, Kluth D, Zander AR, et al. Liver-specific gene expression in mesenchymal stem cells is induced by liver cells. *World J Gastroenterol*. 2005;11(29):4497-504.

[13] Wu S, Ju GQ, Du T, Zhu YJ, Liu GH. Microvesicles Derived from Human Umbilical Cord Wharton's Jelly Mesenchymal Stem Cells Attenuate Bladder Tumor Cell Growth In Vitro and In Vivo. *PLoS One*. 2013;8(4):e61366.

[14] Chapel A, Bertho JM, Bensidhoum M, Fouillard L, Young RG, Frick J, et al. Mesenchymal stem cells home to injured tissues when co-infused with hematopoietic cells to treat a radiation-induced multi-organ failure syndrome. *The journal of gene medicine.* 2003;5(12):1028-38.

[15] Francois S, Bensidhoum M, Mouiseddine M, Mazurier C, Allenet B, Semont A, et al. Local irradiation not only induces homing of human mesenchymal stem cells at exposed sites but promotes their widespread engraftment to multiple organs: a study of their quantitative distribution after irradiation damage. *Stem Cells.* 2006;24(4):1020-9.

[16] Baraniak PR, McDevitt TC. Stem cell paracrine actions and tissue regeneration. *Regen Med.* 2010;5(1):121-43.

[17] De Ugarte DA, Alfonso Z, Zuk PA, Elbarbary A, Zhu M, Ashjian P, et al. Differential expression of stem cell mobilization-associated molecules on multi-lineage cells from adipose tissue and bone marrow. *Immunology letters.* 2003;89(2-3):267-70.

[18] JY O, Kang SK, Choi IS, Ra JC. Comparison of neural cell differentiation of human adipose mesenchymal stem cells derived from young and old ages. *Dev Reprod.* 2009;13:227–37.

[19] Rigotti G, Marchi A, Galie M, Baroni G, Benati D, Krampera M, et al. Clinical treatment of radiotherapy tissue damage by lipoaspirate transplant: a healing process mediated by adipose-derived adult stem cells. *Plastic and reconstructive surgery.* 2007;119(5):1409-22; discussion 23-4.

[20] Phulpin B, Gangloff P, Tran N, Bravetti P, Merlin JL, Dolivet G. Rehabilitation of irradiated head and neck tissues by autologous fat transplantation. *Plastic and reconstructive surgery.* 2009;123(4):1187-97.

[21] Kim SS, Kawamoto HK, Kohan E, Bradley JP. Reconstruction of the irradiated orbit with autogenous fat grafting for improved ocular implant. *Plastic and reconstructive surgery.* 2010;126(1):213-20.

[22] Serra-Renom JM, Munoz-Olmo JL, Serra-Mestre JM. Fat grafting in postmastectomy breast reconstruction with expanders and prostheses in patients who have received radiotherapy: formation of new subcutaneous tissue. *Plastic and reconstructive surgery.* 2010;125(1):12-8.

[23] Salgarello M, Visconti G, Barone-Adesi L. Fat grafting and breast reconstruction with implant: another option for irradiated breast cancer patients. *Plastic and reconstructive surgery.* 2012;129(2):317-29.

[24] Cordeiro PG, Pusic AL, Disa JJ, McCormick B, VanZee K. Irradiation after immediate tissue expander/implant breast reconstruction: outcomes, complications, aesthetic results, and satisfaction among 156 patients. *Plastic and reconstructive surgery.* 2004;113(3):877-81.

[25] Bertho JM, Frick J, Demarquay C, Lauby A, Mathieu E, Dudoignon N, et al. Reinjection of ex vivo-expanded primate bone marrow mononuclear cells strongly reduces radiation-induced aplasia. *Journal of hematotherapy & stem cell research.* 2002;11(3):549-64.

[26] Bertho JM, Bey E, Bottolier-Depois JF, Revel TD, Gourmelon P, J.J. L. *A new therapeutic approach of radiation burns by mesenchymal stem cell transplantation.* Partial-Body Radiation Diagnostic Biomarkers and Medical Management of Radiation Injury Workshop AFFRI, Bethesda, USA. 2008.

[27] Groves AK, Barnett SC, Franklin RJ, Crang AJ, Mayer M, Blakemore WF, et al. Repair of demyelinated lesions by transplantation of purified O-2A progenitor cells. *Nature.* 1993;362(6419):453-5.

[28] Franklin RJ, Bayley SA, Milner R, Ffrench-Constant C, Blakemore WF. Differentiation of the O-2A progenitor cell line CG-4 into oligodendrocytes and astrocytes following transplantation into glia-deficient areas of CNS white matter. *Glia.* 1995;13(1):39-44.

[29] Rezvani M, Birds DA, Hodges H, Hopewell JW, Milledew K, Wilkinson JH. Modification of radiation myelopathy by the transplantation of neural stem cells in the rat. *Radiat Res.* 2001;156(4):408-12.

[30] Acharya MM, Roa DE, Bosch O, Lan ML, Limoli CL. Stem cell transplantation strategies for the restoration of cognitive dysfunction caused by cranial radiotherapy. *J Vis Exp.* 2011(56).

[31] Acharya MM, Christie LA, Lan ML, Giedzinski E, Fike JR, Rosi S, et al. Human neural stem cell transplantation ameliorates radiation-induced cognitive dysfunction. *Cancer Res.* 2011;71(14):4834-45.

[32] Mouiseddine M, Francois S, Souidi M, Chapel A. Intravenous human mesenchymal stem cells transplantation in NOD/SCID mice preserve liver integrity of irradiation damage. *Methods Mol Biol.* 2012;826:179-88.

[33] Xue J, Li X, Lu Y, Gan L, Zhou L, Wang Y, et al. Gene-modified mesenchymal stem cells protect against radiation-induced lung injury. Molecular therapy. *The journal of the American Society of Gene Therapy.* 2013;21(2):456-65.

[34] Semont A, Mouiseddine M, Francois A, Demarquay C, Mathieu N, Chapel A, et al. Mesenchymal stem cells improve small intestinal integrity through regulation of endogenous epithelial cell homeostasis. *Cell Death Differ.* 2010;17(6):952-61.

[35] Linard C, Busson E, Holler V, Strup-Perrot C, Lacave-Lapalun JV, Lhomme B, et al. Repeated autologous bone marrow-derived mesenchymal stem cell injections improve radiation-induced proctitis in pigs. *Stem cells translational medicine.* 2013;2(11): 916-27.

[36] Agay D, Scherthan H, Forcheron F, Grenier N, Herodin F, Meineke V, et al. Multipotent mesenchymal stem cell grafting to treat cutaneous radiation syndrome: development of a new minipig model. *Exp Hematol.* 2010;38(10):945-56.

[37] Francois S, Mouiseddine M, Mathieu N, Semont A, Monti P, Dudoignon N, et al. Human mesenchymal stem cells favour healing of the cutaneous radiation syndrome in a xenogenic transplant model. *Annals of hematology.* 2007;86(1):1-8.

[38] Huang SP, Huang CH, Shyu JF, Lee HS, Chen SG, Chan JY, et al. Promotion of wound healing using adipose-derived stem cells in radiation ulcer of a rat model. *J Biomed Sci.* 2013;20(1):51.

[39] Lombaert IM, Wierenga PK, Kok T, Kampinga HH, deHaan G, Coppes RP. Mobilization of bone marrow stem cells by granulocyte colony-stimulating factor ameliorates radiation-induced damage to salivary glands. *Clin Cancer Res.* 2006;12(6):1804-12.

[40] Puck TT, Marcus PI. Action of x-rays on mammalian cells. *J Exp Med.* 1956; 103(5):653-66.

[41] Herodin F, Mayol JF, Mourcin F, Drouet M. Which place for stem cell therapy in the treatment of acute radiation syndrome? Folia histochemica et cytobiologica / *Polish*

Academy of Sciences, Polish Histochemical and Cytochemical Society. 2005;43(4):223-7.

[42] Caplan AI. Why are MSCs therapeutic? New data: new insight. *J Pathol.* 2009;217(2):318-24.

[43] Coppes RP, Stokman MA. Stem cells and the repair of radiation-induced salivary gland damage. *Oral Dis.* 2011;17(2):143-53.

[44] Rehman J, Li J, Parvathaneni L, Karlsson G, Panchal VR, Temm CJ, et al. Exercise acutely increases circulating endothelial progenitor cells and monocyte-/macrophage-derived angiogenic cells. *Journal of the American College of Cardiology.* 2004;43(12):2314-8.

[45] Lopatina T, Kalinina N, Karagyaur M, Stambolsky D, Rubina K, Revischin A, et al. Adipose-derived stem cells stimulate regeneration of peripheral nerves: BDNF secreted by these cells promotes nerve healing and axon growth de novo. *PLoS One.* 2011;6(3):e17899.

[46] Chen L, Tredget EE, Wu PY, Wu Y. Paracrine factors of mesenchymal stem cells recruit macrophages and endothelial lineage cells and enhance wound healing. *PLoS One.* 2008;3(4):e1886.

[47] Chiellini C, Cochet O, Negroni L, Samson M, Poggi M, Ailhaud G, et al. Characterization of human mesenchymal stem cell secretome at early steps of adipocyte and osteoblast differentiation. *BMC Mol Biol.* 2008;9:26.

[48] Burdon TJ, Paul A, Noiseux N, Prakash S, Shum-Tim D. Bone marrow stem cell derived paracrine factors for regenerative medicine: current perspectives and therapeutic potential. *Bone marrow research.* 2011;2011:207326.

[49] Ratajczak MZ, Kucia M, Jadczyk T, Greco NJ, Wojakowski W, Tendera M, et al. Pivotal role of paracrine effects in stem cell therapies in regenerative medicine: can we translate stem cell-secreted paracrine factors and microvesicles into better therapeutic strategies? *Leukemia.* 2012;26(6):1166-73.

[50] Distler JH, Pisetsky DS, Huber LC, Kalden JR, Gay S, Distler O. Microparticles as regulators of inflammation: novel players of cellular crosstalk in the rheumatic diseases. *Arthritis and rheumatism.* 2005;52(11):3337-48.

[51] Distler JH, Huber LC, Gay S, Distler O, Pisetsky DS. Microparticles as mediators of cellular cross-talk in inflammatory disease. *Autoimmunity.* 2006;39(8):683-90.

[52] Chen TS, Lai RC, Lee MM, Choo AB, Lee CN, Lim SK. Mesenchymal stem cell secretes microparticles enriched in pre-microRNAs. *Nucleic Acids Res.* 2010;38(1):215-24.

[53] Gatti S, Bruno S, Deregibus MC, Sordi A, Cantaluppi V, Tetta C, et al. Microvesicles derived from human adult mesenchymal stem cells protect against ischaemia-reperfusion-induced acute and chronic kidney injury. *Nephrol Dial Transplant.* 2011;26(5):1474-83.

[54] Janowska-Wieczorek A, Majka M, Kijowski J, Baj-Krzyworzeka M, Reca R, Turner AR, et al. Platelet-derived microparticles bind to hematopoietic stem/progenitor cells and enhance their engraftment. *Blood.* 2001;98(10):3143-9.

In: Biological Responses, Monitoring and Protection … ISBN: 978-1-63463-852-4
Editor: K. P. Mishra

Chapter 3

DIFFERENTIAL RADIOSENSITIVITY OF BYSTANDER NORMAL AND TUMOR CELLS AND ITS IMPLICATIONS IN CANCER RADIOTHERAPY

Badri Narain Pandey*

Radiation Biology and Health Sciences Division
Bhabha Atomic Research Centre, Mumbai, India

ABSTRACT

Biological effects of radiation are not only limited to cells exposed to radiation, but radiation damage gets manifested to bystander cells, which have not directly received radiation. This phenomenon of radiation induced bystander effect (RIBE) has been mainly studied in normal cells with relevance to radiation risk assessment. However, bystander effect pertaining to cancer radiotherapy is very limited in literature. Irradiated cancer cells may interact with bystander tumor cells or normal cells through involvement of gap junctions or secretion of diffusible factors. Moreover, normal cells receiving relatively lower doses may also interact with irradiated cancer cells. The magnitude and nature of bystander signaling in tumor microenvironment would have significant implication in clinical outcome of cancer radiotherapy.

INTRODUCTION

Radiation effects are not only limited to cells/organs, which directly receive radiation, but the "biological penumbra" of radiation effects is bigger in bystander cells/organs adjoining to irradiated cells [1-3]. Moreover, there are reports showing abscopal or distant bystander radiation effects [4, 5] exerted in distant tissues/organs after radiation exposure. Bystander effects from irradiated cells are mediated through two major pathways either through gap junctional intercellular communication (GJIC) or through factors secreted from irradiated cells. It may be pertinent to mention here that bystander effects from irradiated cells to

* Tel.: + 91-22-2559 5043/0418; Fax: + 91-22-2550 5151; Email: badrinarain@yahoo.co.in OR bnp@ barc.gov.in.

neighboring cells involve both these mechanisms; however, bystander effects at distant tissues/organs are mediated through factors released from irradiated cells.

Initial radiation induced bystander effect were studied as abscopal effects as early as the 1960's [6, 7] and effects of clastogenic factors in distant organs using irradiated serum samples [8, 9]. However, the major breakthrough in the research area occurred through *in vitro* experiments by Nagasawa and Little [10]. In this study, Chinese hamster ovary cells exposed to α-particles fluences by which only 0.1-1% of the cells' nuclei were actually traversed by a particle track showed sister chromatid exchanges in 20-40% of cells. These results indicated that the target for genetic damage by α-particles is much larger than the nucleus or in fact than the cell itself. These results were confirmed by others for the same endpoint in human diploid fibroblasts [11]. Subsequently, substantial progress has been made in the research area demonstrating RIBE in various experimental models using different end points. Interestingly in one recent study, human lung fibroblasts irradiated to α-particles followed by incubation resulted in higher percentage of cells with γ-H2AX foci, a marker of DNA double strand break, than immediately after radiation (unpublished results). The increased fraction of cells with γ-H2AX foci during course of incubation suggest transmission of bystander signals from alpha irradiated to bystander cells.

	Type	Therapeutic Implications
Tumor to Normal	Toxic	Low
Tumor to Normal	Protective	High
Tumor to Tumor	Toxic	High
Tumor to Tumor	Protective	Low

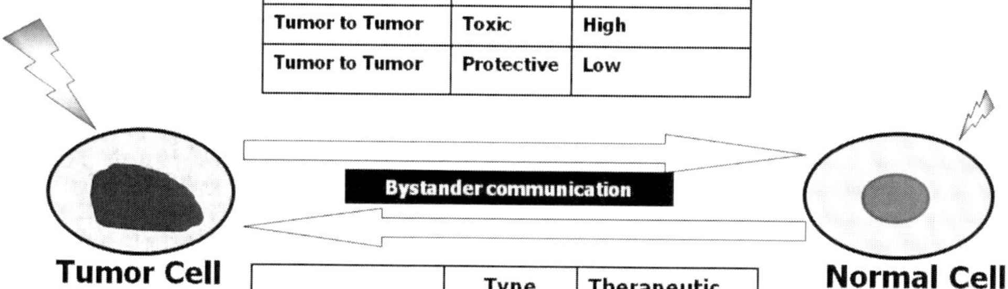

	Type	Therapeutic Implications
Normal to Tumor	Toxic	High
Normal to Tumor	Protective	Low
Normal to Normal	Toxic	Low
Normal to Normal	Protective	High

Figure 1. Cancer Radiotherapy: Bystander Signals and Therapeutic Implications.

Recently, RIBE during cancer radiotherapy scenario has gained attention of radiation biologists to explain the non-targeted effects [12-15] and a few reports showed bystander interaction especially after a high dose of radiation [16]. RIBE holds significant relevance in cancer radiotherapy (Figure 1) due to following reasons: (i) the intermixed boundaries of tumor and normal tissue provide suitable platform for bystander interaction between these cells during radiotherapy. Nevertheless, the magnitude of bystander interaction will vary from these cells as tumor and normal cells generally receive a different range of radiation dose. (ii) During many a clinical scenario, radiation oncologists are unable to irradiate whole tumors either due to lack of diagnosis or due to anatomical restrictions of tumor site. Bystander communication between irradiated tumor to non-irradiated (bystander)/low dose irradiated tumor cells would be expected to play significant role in clinical outcome of radiotherapy in such partial tumor irradiation. (iii) During transformation, most of the tumor cells lose the function of their gap junctions. Malfunctioning of gap junctions is considered, as is one of the reasons of bystander effect mediated cancer chemo- and radio-resistance [17, 18]. (iv) Bystander communication has potential to improve gene therapy based cancer therapy and radiotherapy [19, 20] and various approaches to regulate bystander communication can be exploited to improve the cancer radiotherapy (Figure 2).

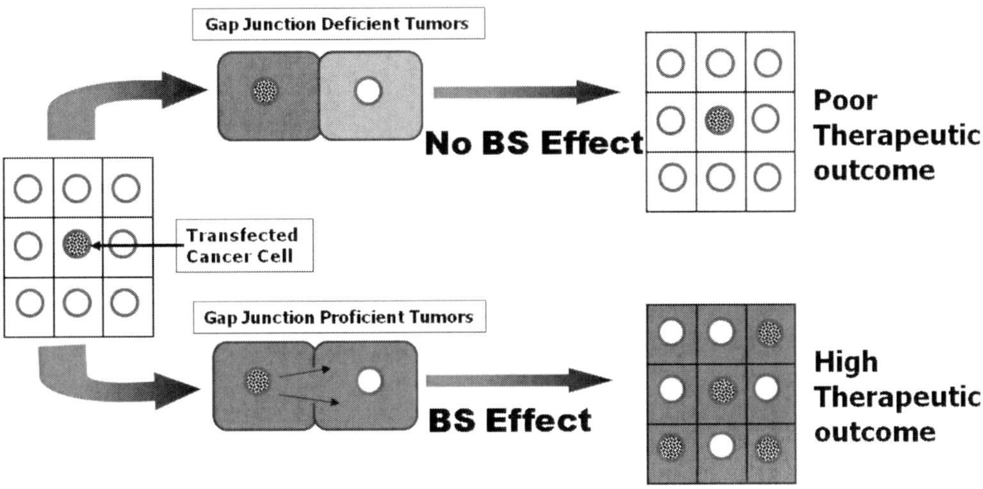

Figure 2. Bystander Response in Cancer Gene Therapy.

It may be very significant to mention that the magnitude and nature (damaging/protective) of bystander crosstalk between irradiated and bystander (or lower dose irradiated) tumor cells would govern whether net bystander effect would enhance the radiation killing to tumor cells or protect them against radiation damage. Bystander effect from irradiated tumor cells to normal cells would determine the short and long term side effects of radiation. On the other hand, the damaging or protective signals from normal cells to counterpart irradiated tumor cells would decide the providence of tumor cells (Figure 1).

RADIOSENSITIVITY OF BYSTANDER CANCER AND NORMAL CELLS

Radiosensitivity of normal and cancer cells in tumor microenvironment is governed by many intrinsic and extrinsic factors like cellular redox level, phase of cell cycle, oxygen concentration etc. [21]. In this regard, there are a few studies showing the radiosensitivity of bystander tumor and cancer cells. Our results showed that irradiated conditioned medium (ICM), obtained from human promyelocytic leukemic cells (HL-60) γ-irradiated from low-dose (5 cGy; ICM-L) and high-dose radiation (1 Gy; ICM-H) showed differential response to bystander (unirradiated) HL-60 and normal human blood lymphocytes. The magnitude of apoptosis in normal lymphocytes cultured in ICM-H was significantly higher than when cells were cultured in ICM-L. Bystander HL-60 cells showed higher sensitivity to undergo apoptosis even after ICM-L treatment. However, ICM-H was found to induce the apoptotic death in both bystander HL-60 and normal lymphocytes [22]. These results suggest differential bystander response from irradiated cancer cells toward cancer and normal cells depending on the dose and/or dose rate of radiation. Such differential bystander response may be associated with fact that the nature and magnitude of factors secreted from tumor cells after low and high dose are different, hence, they may exert different effects to bystander cells.

In another study, we have analyzed the level of cytokines from human tumor cell lines after acute (2, 6Gy) and fractionated doses (3x2Gy) of gamma-irradiation [23]. Irradiated cancer cells showed increase in factors in the culture medium, which resulted in cyto-toxicity to bystander tumor cells. Similar to lower radiosensitivity at dose fractionation, the magnitude of factors secreted from tumor cells was lower in case of dose fractionation than equivalent acute dose suggesting role of these factors in determining the cellular radiosensitivity of bystander cells.

In radiation biology, most of the studies have demonstrated the damaging bystander signaling. However, there have been a few recent studies showing protective bystander signaling, which has relevance in determining the cellular sensitivity of bystander cells. In one such study, it has been shown that dying cells in Drosophila larvae resulted in protection of bystander cells through activation of an anti-apoptotic microRNA, bantam. Moreover, in another finding the reciprocal signal from bystander cells to irradiated cells was studied using α-irradiated human macrophage cells U937 and its bystander HL-7702 hepatocyte cells [24]. Results showed the signals from bystander cells to rescue the irradiated cells through a feedback signal stress involving cAMP and membrane signaling [25].

There are a few recent studies relevant to cancer radiotherapy demonstrating differential radio-sensitivity of cancer and normal cells against radiation induced bystander signals, which were also found to be reciprocal. In one study, irradiated human melanoma (Me45) cells were co-cultured with non-irradiated Me45 cells or normal human dermal fibroblasts (NHDF) cells and *vice versa*. An increase of apoptosis was observed in bystander Me45 cells, which was especially higher in NHDF cells. It was further interesting to observe that the frequencies of micronuclei and apoptosis in irradiated Me45 cells co-cultured with NHDF cells were significantly reduced in comparison with those cultured alone [26].

In our recent study, a direct and precise proton microbeam irradiation system was used to show bidirectional and criss-cross bystander communications between in human lung cancer (A549) and normal (WI38) cells [27]. We observed that in A549–A549 co-cultures, irradiated cancer cells exert damaging effects to bystander A549 cells, which were found to involve GJIC. However, in A549–WI38 co-cultures, irradiated cancer cells did not affect bystander normal cells. It was further interesting to observe that bystander WI38 cells resulted in inverse protective signaling (rescue effect) to irradiated A549 cells; however, this effect was independent of GJIC. It appears that in response to damaging signals from irradiated cells, bystander cells trigger rescue signals to protect themselves, which passively also protects neighboring irradiated cells.

CONCLUSION AND FUTURE DIRECTIONS

Technological advancements like microbeam has made it possible to irradiate cells more precisely with desired dose/number of particles and to elucidate the contribution of various cellular/sub-cellular partners during bystander communication in a population of irradiated-bystander cells. Experimental evidences from various laboratories including ours demonstrated the bystander effects in various experimental systems for different types/doses of radiation. A significant knowledge has been gained about molecular mechanism of radiation mediated bystander effect. However, bystander studies during combinatorial cancer therapy conditions e.g. chemotherapy [28] and further demonstrating the phenomenon using *in vivo* models [29, 30] are limited. More studies in these directions are warranted in future. For better bench-to-bedside translation of knowledge gained from bystander cross-talk studies pertaining to cancer radiotherapy, a closer interdisciplinary feedback based interaction is needed between radio-oncologists and radiation biologists. The better understanding of radiosensitivity of bystander tumor and normal cells would provide two- edge weapon in the battleground of cancer radiotherapy, to save our friends (normal cells) and regulate the bystander signals to effectively kill the enemy (tumor cells).

REFERENCES

[1] Azzam EI, Little JB. The radiation-induced bystander effect: evidence and significance. *Hum Exp Toxicol* 2004;23: 61-5.

[2] Pandey BN, Venkatachalam P, de Toledo SM, Azzam EI. The Ionizing Radiation-Induced Bystander Effect: Mechanisms and Relevance to Radiation Protection and Radiotherapy. *Indian Journal Radiation Research* 2006;2: 13-29.

[3] Suzuki K, Yamashita S. Radiation-Induced Bystander Response: Mechanism and Clinical Implications. *Adv Wound Care (New Rochelle)* 2014;3: 16-24.

[4] Siva S, Macmanus MP, Martin RF, Martin OA. Abscopal effects of radiation therapy: A clinical review for the radiobiologist. *Cancer Lett* 2013.

[5] Jain MR, Li M, Chen W, Liu T, de Toledo SM, Pandey BN, Li H, Rabin BM, Azzam EI. In Vivo Space Radiation-Induced Non-Targeted Responses: Late Effects On Molecular Signaling In Mitochondria. *Curr Mol Pharmacol* 2011;4: 106-14.

[6] Kurnick NB, Nokay N. Abscopal Effects and Bone-Marrow Repopulation in Man and Mouse. *Ann N Y Acad Sci* 1964;114: 528-37.

[7] Law AW, Mole RH. Direct and abscopal effects of x-radiation on the thymus of the weanling rat. *Int J Radiat Biol* 1961;3: 233-48.

[8] Mothersill C, Seymour C. Radiation-induced bystander effects: past history and future directions. *Radiat Res* 2001;155: 759-67.

[9] Vasumathy R, Pandey BN, Mishra KP. Oxidative stress-mediated apoptotic death in human peripheral blood lymphocytes treated with plasma from gamma-irradiated blood. *J Environ Pathol Toxicol Oncol* 2012;31: 1-6.

[10] Nagasawa H, Little JB. Induction of sister chromatid exchanges by extremely low doses of alpha-particles. *Cancer Res.* 1992;52: 6394-6396.

[11] Deshpande A, Goodwin EH, Bailey SM, Marrone BL, Lehnert BE. Alpha-particle-induced sister chromatid exchange in normal human lung fibroblasts - Evidence for an extranuclear target. *Radiat. Res.* 1996;145: 260-267.

[12] Mothersill C, Seymour CB. Radiation-induced bystander effects--implications for cancer. *Nat Rev Cancer* 2004;4: 158-64.

[13] Munro AJ. Bystander effects and their implications for clinical radiotherapy. *J Radiol Prot* 2009;29: A133-42.

[14] Prise KM, O'Sullivan JM. Radiation-induced bystander signaling in cancer therapy. *Nat Rev Cancer* 2009;9: 351-60.

[15] Mothersill CE, Moriarty MJ, Seymour CB. Radiotherapy and the potential exploitation of bystander effects. *Int J Radiat Oncol Biol Phys* 2004;58: 575-9.

[16] Asur R, Butterworth KT, Penagaricano JA, Prise KM, Griffin RJ. *High dose bystander effects in spatially fractionated radiation therapy.* Cancer Lett 2013.

[17] Autsavapromporn N, de Toledo SM, Little JB, Jay-Gerin JP, Harris AL, Azzam EI. The role of gap junction communication and oxidative stress in the propagation of toxic effects among high-dose alpha-particle-irradiated human cells. *Radiat Res* 2011;175: 347-57.

[18] Takano Y, Iwata H, Yano Y, Miyazawa M, Virgona N, Sato H, Ueno K, Yano T. Up-regulation of connexin 32 gene by 5-aza-2'-deoxycytidine enhances vinblastine-induced cytotoxicity in human renal carcinoma cells via the activation of JNK signaling. *Biochem Pharmacol* 2011;80: 463-70.

[19] Cottin S, Gould PV, Cantin L, Caruso M. Gap junctions in human glioblastomas: implications for suicide gene therapy. *Cancer Gene Ther* 2011;18: 674-81.

[20] Kievit E, Nyati MK, Ng E, Stegman LD, Parsels J, Ross BD, Rehemtulla A, Lawrence TS. Yeast cytosine deaminase improves radiosensitization and bystander effect by 5-fluorocytosine of human colorectal cancer xenografts. *Cancer Res* 2000;60: 6649-55.

[21] Hall EJ. Radiation biology for pediatric radiologists. *Pediatr Radiol* 2009;39 Suppl 1: S57-64.

[22] Pandey BN, Kumar A, Ali M, Mishra KP. Bystander effect of conditioned medium from low and high doses of gamma-irradiated human leukemic cells on normal lymphocytes and cancer cells. *J Environ Pathol Toxicol Oncol,* 2011; 30: 333-40.

[23] Desai S, Kumar A, Laskar S, Pandey BN. Cytokine profile of conditioned medium from human tumor cell lines after acute and fractionated doses of gamma radiation and its effect on survival of bystander tumor cells. *Cytokine* 2013;61: 54-62.

[24] Bilak A, Uyetake L, Su TT. Dying cells protect survivors from radiation-induced cell death in Drosophila. *PLoS Genet* 2014;10: e1004220.

[25] He M, Dong C, Xie Y, Li J, Yuan D, Bai Y, Shao C. Reciprocal bystander effect between alpha-irradiated macrophage and hepatocyte is mediated by cAMP through a membrane signaling pathway. *Mutat Res* 2014;763-764C: 1-9.

[26] Widel M, Przybyszewski WM, Cieslar-Pobuda A, Saenko YV, Rzeszowska-Wolny J. Bystander normal human fibroblasts reduce damage response in radiation targeted cancer cells through intercellular ROS level modulation. *Mutat Res* 2012;731: 117-24.

[27] Desai S, Kobayashi A, Konishi T, Oikawa M, Pandey BN. Damaging and protective bystander cross-talk between human lung cancer and normal cells after proton microbeam irradiation. *Mutat Res* 2014;763-764C: 39-44.

[28] Chinnadurai M, Paul SF, Venkatachalam P. The effect of growth architecture on the induction and decay of bleomycin and X-ray-induced bystander response and genomic instability in lung adenocarcinoma cells and blood lymphocytes. *Int J Radiat Biol* 2013;89: 69-78.

[29] Kassis AI. In vivo validation of the bystander effect. Hum Exp Toxicol 2004;23: 71-3.

[30] Xue LY, Butler NJ, Makrigiorgos GM, Adelstein SJ, Kassis AI. Bystander effect produced by radiolabeled tumor cells in vivo. *Proc Natl Acad Sci U S A* 2002;99: 13765-70.

In: Biological Responses, Monitoring and Protection … ISBN: 978-1-63463-852-4
Editor: K. P. Mishra

Chapter 4

Dose Rate: A Critical Factor in Determining the Cancer Risk of Radiation at an Environmental Level

Hiroshi Tanooka,[1,2,*]
[1]Radiation Effects Association, Tokyo, Japan
[2]National Cancer Center Research Institute, Tokyo, Japan

Abstract

The importance of the dose-rate effect in evaluating the cancer risk of ionizing radiation was shown for a wide dose-rate range experienced in the human environment from 10^{-9} Gy/min at the environmental level to 10^{6}-10^{8} Gy/min at the extremely high A-bomb exposure level. The discrepancy between the observed cancer risk and that extrapolated from the A-bomb survivor data with a currently used factor, DDREF, together with the LNT model indicated an overestimation of the risk with the model. Alternately, a non tumor dose, defined as the highest doses at which no significant cancer increase is observed above the control level was applied to analyze the dependence of the cancer risk on the dose rate at various exposure conditions, including dose rate, difference between whole body and partial body exposures, and LET. The value 20 for DDREF at the environmental radiation level compared with the A-bomb radiation level for the whole-body low LET exposure was presented.

Introduction

It was already three years ago when an accident of the nuclear reactor occurred through failure of the cooling system by disconnection of the electric power supply caused by Tsunami on March 11, 2011, in Fukushima, Japan. The areas surrounding the accident site are contaminated with radioactivity, with many problems remaining. One of the most important ones is whether or not the cancer risk in these areas will be increased by the radioactive

* E mail address: tanooka-h@wind.ocn.ne.jp.

contamination. The guidelines for radiation protection rules are based on the severe estimation of the risk according to the recommendation of the International Commission of Radiological Protection (ICRP) [1]. However, overestimation of the radiation risk has caused unnecessary evacuation of residents, unnecessary costs, and most importantly, unnecessary psychological fear. What is needed here is a correct figure for the radiation risk, so that the response to the nuclear accident will be appropriate.

The difference in the cancer risk between a high dose-rate radiation, as in the A-bomb case, and a low dose-rate radiation, as in the present Fukushima case, does not appear to be properly incorporated into the radiation protection rules. ICRP recommended the value 2 for the dose and dose-rate effectiveness factor (DDREF) [2] to extrapolate A-bomb data to the low dose and low dose-rate cases with the linear non threshold (LNT) model. This, however, seems to create an overestimation of the risk as judged from recent evidences from epidemiological and experimental data. Here I present a method for analysis of the dose-rate effect on the radiation cancer risk based on a previous study [3-5] and for determination of a new DDREF value.

1. Range of Dose Rate of Radiation in Human Environment

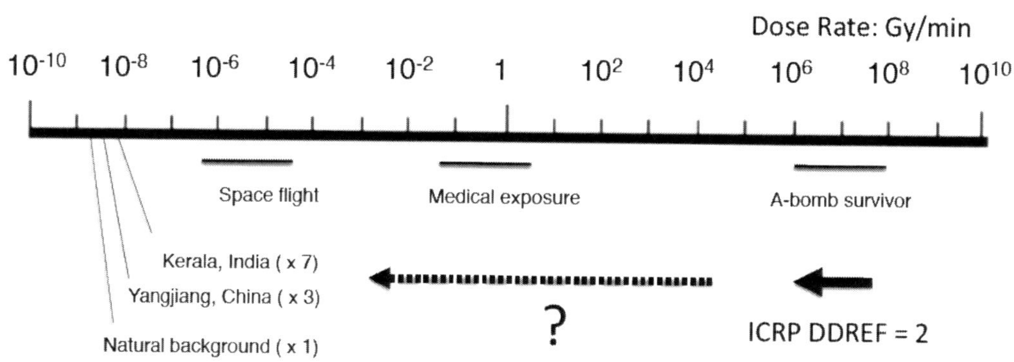

Figure 1. Range of dose rate of ionizing radiation in the human environment. The question of how to apply the dose-rate factor for the radiation cancer risk to the wide dose-rate range is asked.

Humans have been exposed to radiation from the natural environment, medical practice, air travel and, in particular cases, nuclear accidents and nuclear weapons. Each case differs in the dose rate. Figure 1 shows how widely the dose rate is distributed in the human environment from 10^{-9} to 10^{8} Gy/min. An extreme example is the A-bomb survivor cases. The exposure time was assumed to be 1 μsec, which may be longer by other estimates but will not affect the results of analysis, because the dose-rate effect on a cancer risk of radiation appears to saturate at dose rates higher than 1 Gy/min. The lowest dose rate indicated in Figure 1 is in an order of 10^{-9} Gy/min for natural background radiation. Other radiation-exposure cases, such as those of high natural background radiation areas, air travel, space fight, medical diagnosis, radiation therapy, are included in the range between these two extreme dose rates. Internal radiation from radioactivity incorporated in the body is an example of a continuous exposure with a relatively low dose rate and provides valuable information on the dose-rate effect at the median dose-rate range.

Table 1. Radiation dose rates and observed values of non-tumor dose D_{nt} for human cancers

Human case	Exposure condition	Dose rate (Gy/min)	Non-tumor dose, D_{nt}	Data source
A-bomb, leukemia	whole body, low LET	av. 10^{8*}	0.2	[6]
A-bomb, solid cancer	whole body, low LET	av. 10^{8*}	0.1	[7]
Second soft tissue cancer after radiation therapy	partial body, low LET	0.1 – 1	2.5	[8]
Thorotrast, liver cancer	partial body, high LET	1.1×10^{-7}	2	[9]
Radium painter, bone cancer	partial body, high LET	4.9×10^{-7}	10	[10]

To estimate the cancer risk of radiation at various exposure conditions, ideally a set of dose-response curves for cancer incidence at each exposure condition would be needed. However, the number of reports, particularly on humans, is limited. Table 1 shows the available data on humans; leukemia [6] and solid cancers [7] in A-bomb survivors, second cancer after radiation therapy [8], bone cancer in radium dial painters [9], and liver cancer in thorotrast-injected patients [10]. These data show what appears to be a large discrepancy between them. But it indicates a wide variation of human response depending upon the exposure condition. This variation can be explained by taking into consideration the dose-rate effect.

Furthermore, epidemiological data on residents in high natural radiation background areas, such as Yanjiang, China [11] and Kerala, India [12] are very valuable. Their level of γ-radiation from the ground in these areas is 3 and 7 fold higher than in the control areas, respectively. The finding that no cancer increase was detected in these areas answers to the question of how much humans can tolerate an increase of radiation above the natural level.

In an experimental study on the dose-rate problem, the question of how much an increase of the dose rate will or will not increase the cancer risk was quantitatively examined at the Institute for Environmental Sciences, Aomori, Japan. Mice were irradiated continuously with different dose rates of γ rays with daily doses of 0.05, 1, 20 mGy for 22 hours a day for 400 days. Corresponding dose rates were 3.8×10^{-8}, 7.7×10^{-7}, 1.5×10^{-5} Gy/min and total doses delivered were 20, 400, 8000 mGy respectively. Although a problem of this mouse system was the high incidence of spontaneous tumors, very important information was obtained [13]. The increase of cancer risk, expressed by excess % of mice dying from tumors, was 1.1, 2.0, 1.2 for males and -1.5, -2.7, and 1.8 for females respectively. The excess number of primary tumors per mouse was 0.01, 0.05, and 0.36 for males and 0.03, -0.06, and 0.79 (significant increase only in this group) for females respectively. If the ICRP estimation [1], a 0.5% cancer increase per 100 mSv is applied with the LNT model to these mice, the cancer risk increase is estimated to be 0.1, 2, and 40 % for the respective group, showing a large discrepancy from the observation.

2. Limitation of the LNT Model

The LNT model appears to overestimate the cancer risk of radiation. Care should be taken here to see how the value of 2 for DDREF was estimated. ICRP divided the A-bomb survivor data into two groups, i. e., a high dose one and a low dose one. Since the exposure time is the same, the two groups were considered to represent high and low dose-rate groups with a different slope of the linear dose-response lines fitted to the dose-response line of cancer incidence, and then the ratio of the two slopes was obtained as 2 [2]. However, this difference represents the difference in cancer risk between two different dose rates at the extremely high dose-rate region of A-bomb radiation. Even the lowest dose rate for A-bomb survivors is in the order of 10^6 Gy/min, which is 10^{14}-fold higher than the natural background radiation level, 10^{-9}-10^{-8} Gy/min, One possible way to extrapolate DDREF to a lower dose rate of the environmental level is to apply DDREF 2 multiplied by 2^n, where n is the difference in the dose rates between A-bomb and environmental radiation. This estimate will provide a large number of DDREF. However, ICRP provides no answer to this question.

3. Two Historically Important Discoveries in Radiation Biology: Effect of Dose-Rate and Effect of Shielding Part of Body

Upton and coworkers found a decrease of the cancer incidence rate by decreasing the dose rate of X rays [14], employing the RF/Un strain of mice that develop myeloid lymphoma by whole-body X-irradiation. The dose rates employed were in the range from 10^{-4} to 1 Gy/min. This dose-rate effect was further confirmed on other tumors, i. e., thymic lymphoma, ovarian tumors, and lung adenocarcinomas, using the same mouse system with tested dose rates, 0.083/day (3.8 x 10^{-5} Gy/min) vs. 0.4 Gy/min by Ulrich and Storer [15]. Irradiation with neutrons did not show the dose-rate effect within the examined dose-rate range [14]. The finding of the dose-rate effect indicated the existence of a repair mechanism of radiation-induced injuries and was later extended to the study of the central problem of molecular biology, i. e., DNA repair. Certainly, the repair of radiation-induced DNA double strand breaks is dependent on the dose rate and the repair mechanism for DNA base damage is even inducible with a small dose of radiation.

The other important discovery in radiation biology is the protective effect of shielding a part of the mouse body from radiation. Kaplan and Brown found a drastic decrease of the incidence of thymic lymphoma by shielding a half of the body, as compared with the exposure of a whole body in C57BL mice [16]. The shielding of a half part of the mouse body or shielding even one leg protected the mice from death and tumors. This finding indicates the importance of the presence of uninjured blood stem cells in the bone marrow in protecting mice from hematopoietic death and leukemic tumors, possibly involving the immune activity. Kaplan and Brown further developed a regimen for inducing thymic lymphoma with a high frequency; 4 weekly repeats of 1.8 Gy whole-body irradiation yielding 95% tumor incidence [17]. This system was useful also for the study of the dose-rate effect. Ina et al. exposed C57BL mice to γ rays at a very low dose rate, 1.mGy/hr, and compared them with the tumor incidence rate from a 4 acute dose regimen with the same total dose, 7.2 Gy, at 2 Gy/min.

They found no tumors in the mice irradiated at 1 mGy/hr in contrast to the mice exposed to the 4 acute doses that showed 90 % tumor incidence [18]. This result presents an all or none extreme case of the dose-rate effect.

The above findings indicate the importance of the dose rate and the exposure scheme, whole-body or partial-body. These factors should be taken into consideration when accounting for the different results obtained from different exposure conditions. To analyze experimental data together human data obtained in various exposure conditions, a common indicator to express the degree of the sensitivity of cancer induction is needed. A slope of the dose-response curve for cancer induction is one such indicator. However, it can be measured only when the dose-response curve is linear, while in most of cases the dose-response curve for tumor induction is not linear.

4. Definition of Non-Tumor Dose as a Measure of Cancer Risk

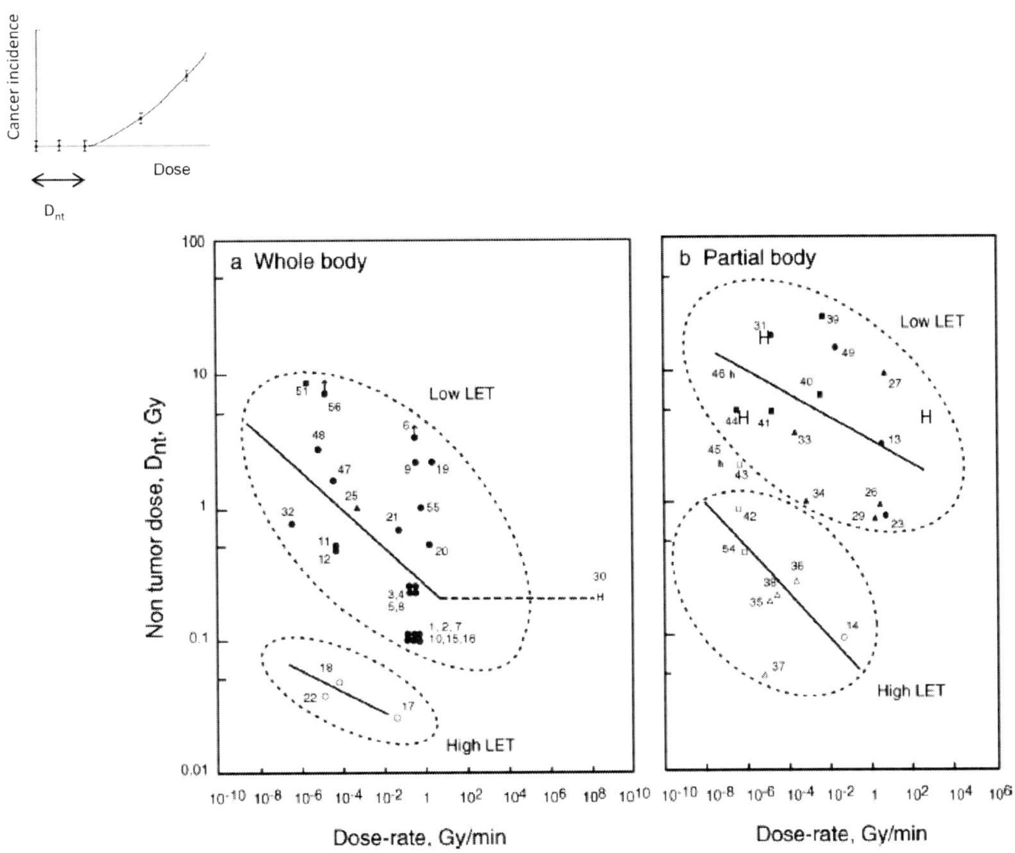

Figure 2. Definition of non-tumor dose, D_{nt} (left upper). Non-tumor dose, D_{nt}, plotted as a function of dose rate for whole body exposure (a) and partial body exposure (b). Experimental data on mice, rats, dogs were combined for analysis. Numbers affixed to each point are the reference number in the previous survey [3, 4]. H: human data. Data collection: see ref. 3 and ref. 4. (Quoted with permission of the publisher.)

The dose-response curve for tumor induction with radiation is not necessarily linear. If the curve is linear, the tumor induction efficiency can be obtained from the slope of the dose-response line and the difference in the tumor-induction efficiency between different radiation-exposure conditions can be obtained by comparison of the slopes. However, the dose-response curve skews depending on radiation-exposure conditions such as dose rate of radiation, whole-body and partial body exposure, and LET of radiation. In these cases, an indicator other than the slope of the dose-response curve is needed to express the tumor induction efficiency. As a measure, the UNCEAR report summarized the minimum doses to induce tumors by radiation surveyed in the literature [19]. However, this measure misses the low-dose region in the dose-response curve, when the data were limited in the high dose region.

Alternatively, the highest dose of radiation at which no statistically significant increase of tumor incidence was observed above the control level was defined as non tumor dose (D_{nt}) [3, 4]. D_{nt} can be an indicator of the tolerance level for cancer induction or the safety level for radiation-cancer risk. (Figure 2, left upper).

5. Survey and Analysis of Non-Tumor Doses

D_{nt} values for five human cases are listed in Table 1. The values were further surveyed on experimental animals including mice, rats, and dogs, in the literature [3, 4]. The number of informative reports showing the dose-response data of tumor incidence, particularly at low doses, is limited. Those data include external and internal irradiation cases, where the external radiation cases provide a more accurate value of the dose and dose rate. The internal radiation cases required calculations to obtain an average dose rate over the decay period of radioactivity in the target organ or whole body. However, the internal radiation data provided valuable information for low dose rates below 1 Gy/min down to the environmental level.

The numerical values of D_{nt} for about 50 cases surveyed in the literature were shown in previous reports [3, 4]. The plotting of D_{nt} as a function of the dose rate of radiation on the log-log scale is shown in Figure 2. These D_{nt} data were divided into four categories classified by the exposure conditions. Equations for the regression lines fitted to the four groups are as follows:

$$D_{nt} = 2.69Dr^{-0.0857} \text{ for low LET partial-body} \qquad \text{(eq. 1)}$$

$$D_{nt} = 0.258Dr^{-0.141} \text{ for low LET whole body} \qquad \text{(eq. 2)}$$

$$D_{nt} = 0.0439Dr^{-0.167} \text{ for high LET partial body} \qquad \text{(eq. 3)}$$

$$D_{nt} = 0.0207Dr^{-0.0733} \text{ for high LET whole body} \qquad \text{(eq. 4)},$$

where Dr is the dose rate of radiation and an applicable dose-rate range for these equations is from 10^{-9} to 1 Gy/min (Figure 3). The dose-rate effect for the whole-body low-LET exposure appeared to saturate above 1 Gy/min. Therefore, the DDREF value 2 in the extremely high dose-rate range of A-bomb radiation [2] is questionable.

High LET radiation has been thought to exert a no dose-rate effect. This problem is important especially in assessing the cancer risk in astronauts, uranium miners, or even in residents in a house where radon is found. Because of a putative high RBE (relative biological effectiveness), a high LET radiation is thought to be particularly harmful. However, the presence of the dose-rate effect was discovered for a very low dose-rate of radon α-rays on the lung carcinoma induction in rats, where the relative risk of radon administered at 150 WL (1.5×10^{-4} Gy/min according to the author's calculation) was diminished to the control level when the dose rate was decreased to 2 WL (2×10^{-6} Gy/min) with the same total dose, 25 WLM (0.19 Gy)[20]. This dose-rate effect was reconfirmed in the repeated experiments [21]. Therefore, the presence of the dose-rate effect for high LET radiation exposure to the partial-body, the lung in this case, is undoubted in contrast to what was previously thought.

From above considerations, an apparent discrepancy in the cancer risk of radiation, even 1000 fold difference in the risk such as that between the A-bomb case and the radium painter case, can be easily visualized from Figure 3.

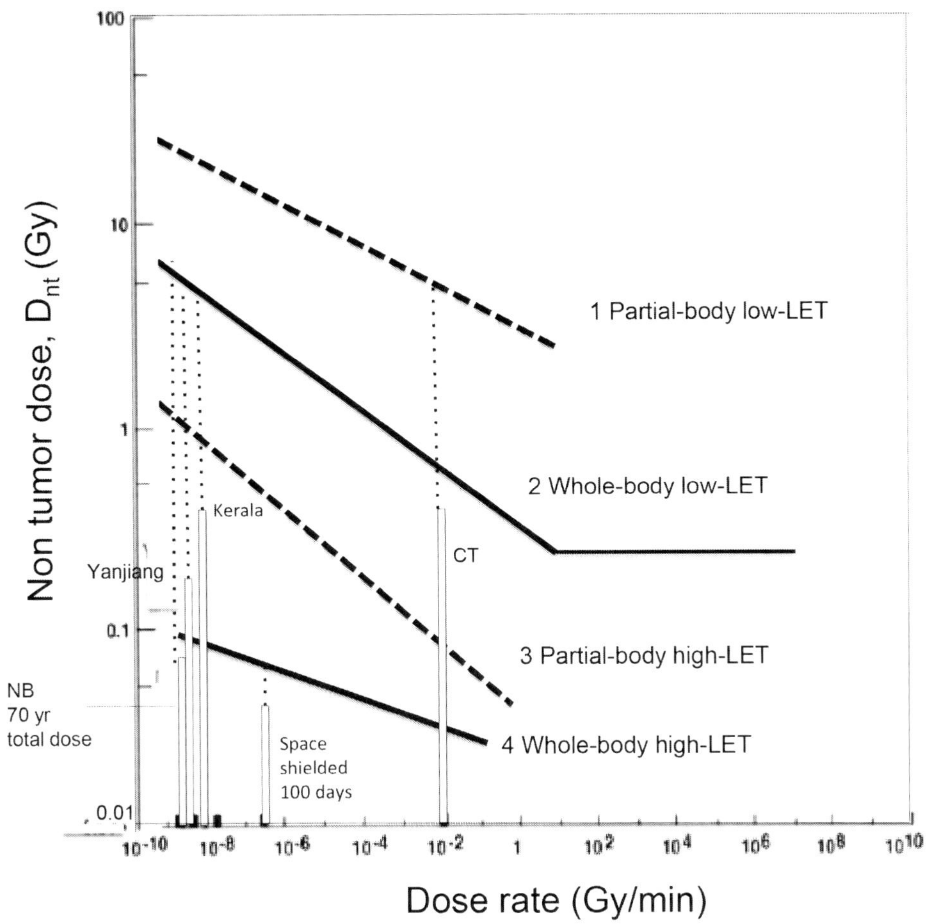

Figure 3. Regression lines fitted to 4 groups: a. Low LET partial body exposure. b. Low LET whole body exposure. c. High LET partial body exposure. d. High LET whole body exposure. Vertical bars indicate total radiation doses received by residents in natural or high radiation background areas for 70 years, space travelers for 100 days, and CT–scanned patients. (Quoted with permission of the publisher.)

6. Application of Non-Tumor Doses to Estimation of Tolerance Level for Cancer Risk

The question of how much the elevation of the natural background radiation level can be tolerated for the cancer risk to humans was quantitatively examined by using the D_{nt} values. The natural background radiation is a complex of the whole-body low-LET and the partial-body high-LET radiations. Radon 222 and its progeny in the natural environment emit α rays and its target organ is the lung. In this respect, radon exerts a partial-body high-LET radiation. Radon contributes 1.3 mSv or 0.13 mGy per year in addition to other radiation sources for whole-body low-LET radiations; cosmic ray (0.38 mGy), internal radiation (0.24 mGy), and ground γ rays (0.46 mGy) [22]. The idea of the non tumor dose was applied to estimate the tolerance level for the increase of the cancer risk by elevation of the natural background radiation level treated as the whole-body low-LET exposure, including the cases in Yanjiang, China (x 3 increase) and Kerala, India (x 7 increase). Table 2 shows the total lifetime (70 years assumed) doses compared with the corresponding values of D_{nt}. The ratio of D_{nt} to the total dose is an index for the cancer risk and defined as the cancer risk index (CRI). CRI 1 means that the total dose is at the lowest edge of cancer risk, since D_{nt} is the highest dose for the absence of cancer increase. It can be seen that CRI is far below 1 in Yanjian and Kerala [Figure 3. Table 1] and exceeds 1 when the γ radiation background becomes 100 times higher than the natural level. This calculation can be extended further to estimation of the cancer risk of radiation in other cases such as space travel, radon increase, medical exposure, and accidental exposure.

Table 2. Increase of the natural background γ radiation level, corresponding total lifetime dose, non tumor dose D_{nt}., and cancer risk index

Increase of background γ radiation[a]		Dose rate (Gy/min)	Total dose, 70 years (Gy)	D_{nt} calculated[b] (Gy)	Cancer risk index, CRI (Total dose/D_{nt})
x 1		2.30×10^{-9}	0.085	4.4	0.020
x 3	(Yanjiang)	4.05×10^{-9}	0.150	4.0	0.038
x 7	(Kerala)	7.55×10^{-9}	0.278	3.6	0.077
x 10		1.03×10^{-8}	0.378	3.5	0.110
x 100		8.89×10^{-8}	3.27	2.5	1.29
x 1000		8.77×10^{-7}	32.3	1.8	17.5

[a]Natural γ radiation from the ground: 0.46 mGy/yr out of the total dose, 1.21 mGy/yr [22]. Increase of this component was considered in calculation.

[b]Equation 2 for whole-body low-LET exposure was applied.

The cancer risk of radioactive cesium and iodine is a current problem of estimating health risk after the nuclear accident. These risks are treated in the category of the whole-body low-LET and the partial-body low-LET exposure, respectively. Space travel is a case of whole-body high-LET radiation exposure. The shielding is not necessarily protect astronauts from high LET galactic cosmic rays, because it creates even more high LET particles and scattered radiation by nuclear collision [23]. The total dose for a 100-day travel in the space becomes close to the corresponding D_{nt} of the whole-body high-LET exposure, if all radiation in a shielded condition is assumed to be with high LET. Furthermore, the radiation dose for the 30 month Mars mission is nearly 1 Sv [23]. On the other hand, diagnostic radiation of computer tomography (CT) is a case of partial body exposure with low LET. Although the considerably high cancer risk of CT was estimated from the LNT model with the value of DDREF, 2 [24], the CT-radiation dose is far below the corresponding D_{nt} (Figure 3) and therefore, no cancer risk is expected from this D_{nt} value.

Further development of the D_{nt} method will be needed in its application to the mixed radiation exposure that exerts different carcinogenic effects at different target organs.

CONCLUSION

A new method for estimating the cancer risk of ionizing radiation at the environmental level and under other exposure conditions was demonstrated using the inverse relation between dose rate and non-tumor dose D_{nt}. The D_{nt} value varies in a wide range of magnitude and explains a large discrepancy between cancer frequencies in previously reported data. The difference in the non tumor dose between environmental radiation and A-bomb radiation was 20 in the ratio, indicating a much higher DDREF value compared with the currently applied value of DDREF, 2. Humans, it appears, are much more tolerant to radiation than has been thought.

ACKNOWLEDGMENTS

I wish to thank Dr. Gerda Horneck and Dr. Thomas Berger, DLR German Aerospace Center, for valuable information on the estimates of the radiation dose from galactic cosmic rays.

REFERENCES

[1] ICRP. (2006). Low-dose extrapolation of radiation related cancer risk. Publication 99. Oxford, Elsevier.

[2] ICRP. (1991) 1990-Recommendation of the International Commission on Radiological Protection. Publication 60. Annals of the ICRP. Oxford, Pergamon Press.

[3] Tanooka, H. (2001). Threshold dose-response in radiation carcinogenesis: An approach from choronic β-irradiation experiments and a review of non-tumor doses. *Intern.J. Rad. Biol.*, 77, 541-551.

[4] Tanooka, H. (2011). Meta-analysis of non-tumour doses for radiation-induced cancer on the basis of dose-rate. *Intern. J. Rad. Biol.*, 87, 645-652.

[5] Tanooka, H. (2011) Dose rate dependence of radiation cancer risk as measured by non-tumor dose. *Health Phys.*, 100, 304 -305.

[6] Shimizu, Y., Kato, H. & Schull, W. J. (1990). Studies of the mortality of A-bomb survivors. 9. Mortality, 1950-1985: Part 2: Cancer mortality based on the recently revised doses (DS86). *Radiat. Res.,* 121, 120-140.

[7] Pierce, D. A. & Preston, D. L. (2000). Radiation-related cancer risks at low doses among atomic bomb survivors. *Radiat. Res.*, 154, 178-186.

[8] Tubiana, M., Diallo, I., Chavaudra, J., Lefkopoulos, D., Bourhis, J., Girinsky, T., Brider, A., Hawkins, M., Haddy, N., El-Fayech, C., Adjadj, E., Clero, E. & de Vathaire, F. (2011). A new method of assessing the dose-carcinogenic effect relationship in patients exposed to ionizing radiation. A concise presentation of preliminary data. *Health Phys..*, 100, 296-299.

[9] Rowland, R. E., Stehney, A. F., & Lucus, H. F. (1978). Dose-response relationships for female radium dial workers. *Radiat. Res.*, 76, 368-383.

[10] Anderson, M. & Storm, H. H. (1992). Cancer incidence among Danish thorotrast-exposed patients. J. *Nat. Cancer Inst.*, 84, 1318-1325.

[11] Chen, D. & Wei, L. X. (1991). Chromosome aberration, cancer mortality and hormetic phenomena among inhabitants in areas of high background radiation in China. *J. Radiat. Res.*, 32 suppl. 2, 46-53.

[12] Nair, M. K., Nambi, K. S. V., Amma, N. S., Gangadharan, P., Jayalekshmi, P., Jayadevan, S., Cherian, V. & Reghuram, K. N. (1999). Population study in the high natural background radiation area in Kerala, India. *Radiat. Res.*, 152, S145-148.

[13] Tanaka, I. B. III, Tanaka, K., Ichinohe, K., Matsumoto, T., Otsu, H., Ogiso, Y. & Sato, F. (2007). Cause of death and neoplasia in mice continuously exposed to very low dose rates of gamma rays. *Radiat. Res.*, 167, 417-437.

[14] Upton, A. C., Randolph, M. L., and Conklin, J. W. (1970). Late effects of fast neutrons and gamma-rays in mice as influenced by the dose rate of irradiation: Induction of neoplasia. *Radiat. Res.*, 41, 467-491.

[15] Ullrich, R. L. & Storer, J. B. (1979). Influence of γ irradiation on the development of neoplastic disease in mice. I. Reticular tissue tumors. *Radiat. Res.*, 80, 303-316.

[16] Kaplan, H. S. & Brown, M. B. (1951). Further observation in inhibition of lymphoid tumor development by shielding and partial-body irradiation of mice. *J. Natl. Cancer. Inst.*, 12, 427-436.

[17] Kaplan, H. S. & Brown, M. B. (1952). A quantitative dose-response study of lymphoid-tumor development in irradiated C57 black mice. *J. Natl. Cancer Inst.* 13, 185-208.

[18] Ina, Y., Tanooka, H., Yamada, T. & Sakai, K. (2005). Suppression of thymic lymphoma induction by life-long low dose-rate irradiation accompanied by immune activation in C57BL/6 mice. *Radiation Research.*, 163, 153-158.

[19] United Nations Scientific Committee on the Effects of Atomic Radiation (UNSCEAR). (2000). Biological effects at low radiation doses: Models, mechanisms and uncertainties. In: Sources and Effects of Ionizing Radiatjion: Genetic and Somatic Effects of Ionizing Radiation. Report to the Genral Assembly, Annex I. New York, United Nations.

[20] Morlier, J. P., Morin, M., Monchaux, G., Fritsch, P. Pineau, J. F., Chameaud, J., Lafuma, J. & Masse, R. (1994). Lung cancer incidence after exposure of rats to low doses of radon: Influence of dose-rate. *Radiat. Protect. Dosimet.*, 56, 93-97.

[21] Monchaux, G, Morlier, J. P., Altmeyer, S., Debroche, M. & Morin, M. (1999). Influence of exposure rate on lung cancer induction in rats exposed to radon progeny. *Radiat. Res.*, 152, S137-S140.

[22] United Nations Scientific Committee on the Effects of Atomic Radiation. (1993). Sources and Effects of Ionizing Radiation. UNCEAR 1993 Report to the General Assembly, with Scientific Annexes. New York, United Nations.

[23] Durante, M. & Cucinota, F. A. (2011). Physical basis of radiation protection in space travel. *Rev. Mod. Phys*. 83, 1245-1281.

[24] Berrinton de Gonzalez, A. & Darby, S. (2004). Risk of cancer from diagnostic X-rays: Estimates for the UK and 14 other countries. *Lancet*, 363, 345-361.

In: Biological Responses, Monitoring and Protection … ISBN: 978-1-63463-852-4
Editor: K. P. Mishra

Chapter 5

RADIATION RISK FROM MEDICAL X-RAY DEVICES, INTERVENTIONAL RADIOLOGY AND MAMMOGRAPHY

***Dr. S. P. Mishra*[1] *and Prof M. L. B. Bhatt*[2]**
[1]Sr. Radiation Scientist, Department of Radiation Oncology,
[2]Prof of Radiation Oncology
Dr. Ram Manohar Lohia Institute of Medical Sciences,
Gomti Nagar, Lucknow, Uttar Pradesh, India

ABSTRACT

Risk and benefit analysis of radiation have engaged the mind of scientific community from several decades. The basic quantification of possible consequences from radiation exposure were estimated from horrendous consequences of Japanese Atomic Bomb survivors. Ionizing radiations and their exposure to human being have the possibility of producing various degrees of harmful effects depending upon the quality and severity of the radiation. Therefore, investigations using ionizing radiation have to be justified by a clear benefit to the patient. In this article a review of quantitative aspects of exposures emanating from conventional X-rays, Multi Slice CT scanners, Interventional radiology practices and mammography have been summarized. The radiation doses from CT scan study comprising of 20 slices may be equivalent to 75- 150 Chest X-ray dose to a single patients. We have evaluated the organ dose and ESD (Entrance Surface Dose) emanating from various CT procedures. Well calibrated thermoluminescence dosimeters (TLD) were utilized for measurement of doses. CT as a modality has found new applications in angiography and is being increasingly used, encompassing up to 40% of all radiographic studies. The mean (± SD) skin doses varied from 728 ± 382, to 128 ± 81, mGy, and the mean effective doses varied from 119.7 ± 50.3, to 9.1 ± 5.5 mSv for some of the most performed angiography procedures. The maximum skin dose was 1.95 Gy, in one of the studies. Mammography is an extremely useful non-invasive imaging technique with inbuilt advantages for the detection of early breast cancer but doses to glandular tissues impose risk of induction of malignancy. The doses received by the patients were found to be dependent on compressed breast thickness, for breast cancer patients (0.8 - 3.8 mGy for 7.9 cm to 3.8 mGy for 11.6 cm) and fibrocystic disease / breast cystic (0.3 m

Gy for 5.5 cm to 1.6 mGy for 6.2 cm). Some of the somatic effects of radiation depend on the severity of radiation doses which may manifest as sterility (temporary or permanent), loss of appetite, impairment of vision (cataract), leukemia, blackening of skin, shortening of life span, carcinogenesis etc. The article will elaborate its pertinent details. The short-term risk to patients is radiation-induced skin damage in cardiac angiography, which can result from acute radiation doses of $\geq$2Gy is a matter of concern and challenge to develop such technologies which are safe and accurate. The extent of the skin injury may not be known for weeks after the procedure. Repeated procedures increase the risk of skin injury, because previous radiation exposure sensitizes the skin. Long term effects include the potential risk of cancer. It is generally accepted that there is probably no low dose "threshold" for inducing cancers, i.e. no amount of radiation should be considered absolutely safe. Current article attempts to summarize exposure in diagnostic practices with possible health consequences.

INTRODUCTION

The introduction of X-ray and other ionizing radiation for clinical applications has significant beneficial effects in the management of human ailments. With passage of time, availability of X-ray based diagnostic aids, the frequency and total population exposer of X-ray radiation are expected to increase several fold. All ionizing radiation and their exposure to human beings have the possibility of producing various degrees of harmful effects depending upon the quality and severity of the radiation.

Risk and benefit analysis of radiation has engaged the minds of the scientific community for several decades. The basic framework of radiation hazard emerged from horrendous consequences of Japanese Atomic bomb survivors. Since then the concern for safety has been discussed in several International Commission on Radiation Protection (ICRP) reports. WHO has identified radiation safety from ionizing radiation exposures as one of its priority agenda for evaluation and reduction of organ doses to the patients in diagnostic radiology. It has been generally agreed that the unnecessary high doses from diagnostic X-ray applications are due to investigations often requested without adequate evaluation of the patient and consideration for use of alternative techniques. Some of the cusaetive factors for unwarranted dose in diagnostic radiology may be attributable to high repetition rate which may not be justified, incorrect radiographic techniques, inappropriate or poorly functional equipment, poorly trained personnel and unplanned installation of the x-ray and CT facilities. The absent legal frame work and nonexistent practices of periodic accreditation has contributed to this avoidable radiation burden. The diagnosis of diseases using conventional X-ray, CT scanner, angiography, Mammography, and PET-CT have increased multiple folds. The tendency to prescribe top end procedure as the first step has been fuelled by speculative superiority of top end modality as reflection of concern and lack of awareness and non-evidence based use of such investigations.

Therefore, investigations using ionizing radiation have to be justified by a clear benefit to the patient. The radiation dose from CT scan study comprising of 20 slices may be equivalent to 75- 150 chest X-ray dose to a single patient. Mammography is an extremely useful non-invasive imaging technique for the detection of early breast cancer. It plays an important role in the screening of women above a certain age or with a family history of breast cancer. The Mammography with 3D capability (tomo-synthesis) has opened a new vista but has the high dose constraint and needs judicious applications. The angiography study based on X-ray C-arm or CT scan has found new vigor both in diagnostic and therapeutic applications. It has changed lifesaving interventional approach in cardiology, neurology, orthopedics, pediatrics and many other super specialties. Nevertheless, a fundamental requirement of the International Basic Safety Standards for Protection Against Ionizing Radiation (BSS) is the optimization of radiological protection of patients undergoing medical exposure and this needs to be addressed with front line knowledge without compromising the patient care.

The current scientific knowledge indicates that even the low doses of ionizing radiation have ability to enhance the cancer risk several folds than what was estimated earlier. The newer studies indicate that low level of ionizing radiation have the potential to affect the I.Q. of the unborn if it was exposed in the womb. Patient has the right to be explained about the possible beneficial and harmful effects from the diagnostic procedures. It is possible to explain the risk estimates (both genetic and somatic) to a patient undergoing barium meal, IVP or any abdomino-pelvic, thorax, head CT or X-ray investigation and interventional procedures. This article attempts to provide an over view of radiation risk from medical x-ray devices, interventional radiology, mammography and other diagnostic procedures.

Radiation Dose Levels for Interventional CT Procedures

In India about 50000 x-ray units are at work in various clinics. The number of CT scanners and more recently PET-CT have seen the geometric rise. It is estimated that about 8000 CT scanners are at work in the country and number of PET-CT centers will be touching 500 by the end of 2014. In the USA alone about 62 million CT scans were performed in the year 2006 as compared to just three million in 1980. There are good clinical reasons for this trend but it also demands judicious use and caution in such applications. CT scanning and nuclear imaging have revolutionized diagnosis and treatment of various disease conditions and have almost eliminated the need for once-common exploratory surgeries and many other invasive and potentially risky procedures. The benefit of these tests, when they're appropriate, far outweigh any radiation-associated cancer risk, whereas the risk from a single CT scan or nuclear imaging test is quite small. In the light of about 20-fold increase in the use of these tests in recent times. Experts wonder if we are risking future public health.

In April 2010 a report was published about "Reducing Environmental Cancer Risk: What We Can Do Now." Among other concerns, the report highlighted the rise in radiation exposure from medical imaging as one of the most important factor for public health. It suggested ways to minimize radiation exposure from medical sources and recommended that clinicians keep a tally of the amount of radiation their patients receive from medical imaging.

Exposure to Ionizing Radiation on the Rise

The radiation we get from medical imaging, viz., X-ray, CT, nuclear imaging and PET-CT procedure penetrate the deep tissue inside body to reveal the internal organs and structures. Ionizing radiation can damage DNA, although the body cells repair most of the damage, but it may leave a small area of "mis-repair." The result of this "mis-repair" in DNA mutation may contribute to development of cancer several years later.

Humans have been exposed to small doses of ionizing radiation from natural sources all the time in particular, cosmic radiation, mainly from the sun and radon, a radioactive gas that comes from the natural breakdown of uranium in soil, rock, water, and building materials. How much of this background radiation one gets exposed, depends on many factors, including altitude and home ventilation. But this may be as high as 3 millisieverts (mSv) per year. Exposure to ionizing radiation from natural or background sources has not changed since about 1980. But average individual radiation exposure has nearly doubled, in USA and reasons attributed is increased use of medical imaging. The proportion of total radiation exposure that comes from medical sources has grown from 15% in the early 1980s to 50% today. CT alone accounts for 24% of all radiation exposure in the United States, according to a report issued in March 2009 by the National Council on Radiation Protection and Measurements.

Table 1. Radiation Doses of Medical Imaging Procedures

Modality	Site /organ	Dose Range, mSv	Average Dose, mSv	Chest X-ray Equivalent Dose
X-rays				
	Chest	0.02-0.67	0.34	1
	C-spine	0.063-0.27	0.17	0.5
	T-spine	0.4-1.4	0.9	2.6
	L-spine	0.8-2.4	1.6	4.7
	Pelvis	0.7-0.86	0.78	2.3
	Abdomen, kidneys, ureters, bladder	0.5-1	0.75	2.2
	Hip	0.3-0.6	0.4	1.1
	Limbs	0.01-0.06	0.035	0.1
	Barium enema	7-9	8	23.5
	Intravenous pyelogram (IVP)	2.5-5.7	4.1	12
	Mammography	0.07-0.89	0.48	1.4
	Upper GI tract	3.6	3.6	10.6
	Dental	0.02-0.334	0.18	0.53
CT scans				
	Head	1.5-2.3	1.9	5.6
	Chest	4.1-8	6	17.6
	Thoracic	8.3-11.7	10	29.4
	Lumbar	3.5-5.2	4.4	13
	Abdominal	7.6-16	11.8	35
	Pelvis	10-13	11.5	33.8
Angiographs				
	Cerebral	7.5	7.5	22
	Cardiac	71.9	71.9	211.5
	Vascular	19.4	19.4	57

Authors have evaluated the organ dose and ESD (Entrance Surface Dose) emanating from various CT procedures. Well calibrated Thermo-Luminiscence Dosimeters(TLD) were utilized for measurement of doses. CT as a modality has found new application in angiography and as such has been increasingly being used, encompassing up to 40% of all radiographic studies. We performed study to determine average radiation dose level to patients undergoing CT-guided interventional procedures. A total of 595 patients undergoing CT interventional procedures were included in this study. Enrolled patients underwent one of five procedures viz aspiration, biopsy, cardiac carotid angiogram, brain angiogram and scans were done in helical mode. Skin dose was estimated from the volumetric CT dose index (CTDI(vol)) and phantom measurements were also taken for calibration. Effective dose was calculated by multiplying dose-length product (DLP) and conversion factor (k factor) for helical mode, and using Monte Carlo organ dose coefficients for intermittent mode. The mean (± SD) skin doses were 728 ± 382, 130 ± 104, 128 ± 81, 152 ± 105, and 195 ± 147 mGy, and the mean effective doses were 119.7 ± 50.3, 20.1 ± 11.0, 13.8 ± 9.2, 25.3 ± 15.4, and 9.1 ± 5.5 mSv for each of the five procedures, respectively. The maximum skin dose was 1.95 Gy. The mean effective dose across all procedure types was 24.1 mSv, from helical scan. Substantial dose differences were observed among the five procedures. The risk of deterministic effects appear to be very low, because the maximum observed skin dose did not exceed the threshold for transient skin erythema (2 Gy). The average risk of stochastic effects were up to a maximum of 10 times of an average CT examinations. Summary of the measurements of CT abdomen are pelvis are detailed in Table 1 with comparisons.

Ionizing Radiation and Cancer Risk

Radiation is utilized both for cancer treatment and diagnosis. It has the potential to cure cancer and induce cancer. Children and teens who receive high doses of radiation to treat lymphoma or bone tumors, specially Ewing's sarcoma and other cancers are more likely to develop additional cancers later in life. It has been possible to have retrospective evaluation but the subject matter does not permit to undertake clinical trials about cancer risk from medical radiation in healthy adults. The knowledge is based on animal data or deductions from long term surviving cancer patients having second malignancy. The risk estimation of ionizing radiation comes from long-term studies of people who survived the 1945 atomic bomb blasts at Hiroshima and Nagasaki in Japan.

These studies show a slightly but significantly increased risk of cancer in those exposed to the blasts, including a group of 25,000 Hiroshima survivors who received less than 50 mSv of radiation — an amount one might get from two or three CT scan examination. The United Nations Scientific Committee on Effect of Atomic Radiation (UNSCEAR) report 1985, highlights that the risk estimated from low level exposure of ionizing radiation is highly under estimated and this sent a ripple all over the world and ICRP recommended $1/^{3rd}$ reduction of maximum permissible dose for radiation workers and for general public it was reduced from$1/10^{th}$ to $1/20^{th}$ of radiation dose to radiation workers The atomic bomb survivors is not a perfect model for exposure to medical radiation, because the bomb released its radiation all at once, while the doses from medical imaging are smaller and spread over time. Still, most experts believe that low level diagnostic radiation can be almost as harmful as getting an equivalent dose in acute exposure Some of the other authors have compiled the data of exposure as detailed in Table 2. (Source: Mettler FA, et al. "Effective Doses in Radiology and Diagnostic Nuclear Medicine: A Catalog," *Radiology* (July 2008), Vol. 248, pp. 254–263).

Table 2. Imaging procedures and their approximate effective radiation doses*

Procedure	Average effective dose (mSv)	Range reported in the literature (mSv)
Bone density test+	0.001	0.00–0.035
X-ray, arm or leg	0.001	0.0002–0.1
X-ray, panoramic dental	0.01	0.007–0.09
X-ray, chest	0.1	0.05–0.24
X-ray, abdominal	0.7	0.04–1.1
Mammogram	0.4	0.10–0.6
X-ray, lumbar spine	1.5	0.5–1.8
CT, head	2	0.9–4
CT, cardiac for calcium scoring	3	1.0–12
Nuclear imaging, bone scan	6.3	7-12
CT, spine	6	1.5–10
CT, pelvis	6	3.3–10
CT, chest	7	4.0–18
CT, abdomen	8	3.5–25
CT, colonoscopy	10	4.0–13.2
CT, angiogram	16	5.0–32
CT, whole body	variable	20 or more
Nuclear imaging, cardiac stress test	40.7	40-50

*The actual radiation exposure depends on many things, including the device itself, the duration of the scan, body size, and the sensitivity of the tissue being targeted.

+Dual energy x-ray absorptiometry, or DXA.

In a 2009 study from Brigham and Women's Hospital in Boston, researchers estimated the potential risk of cancer from CT scans in 31,462 patients over 22 years. For the group as a whole, the increase in risk was found to be 0.7% above the overall lifetime risk of cancer in the United States, which is 42% due to all possible risk. But for patients who had multiple CT scan examination, the increase in the risk was higher, which ranged from 2.7% to 12%. (in this group, 33% participants had received more than five CT scans; 5%, more than 22 scans; and 1%, underwent more than 38 CT scan examination)

Patient need to discuss any high-dose diagnostic imaging: CT or nuclear scan are required to treat or diagnose a medical condition, the benefits usually outweigh the risks. Patient has the right to know the possible life time risk, radiation doses and possible alternative means to make a choice. It is generally recommended by ICRP/IEC/ICRU that imaging device makers indicate the radiation dose for each x-ray, and that clinicians record radiation exposures in patients' medical records. If a clinician recommends a CT or nuclear medicine scan, patient can ask if another non-ionizing radiation technique would suffice, such as an ultrasound (which uses high-frequency sound waves) or MRI (which relies on magnetic energy). It may be important to consider less-frequent testing without compromising the quality of life. If a patient is required to undergo regular CT scans for a chronic condition, it may be explored whether it is possible to increase the time interval between scans. Patient should never seek out CT scan on their own. *Public may in a general be educated about not to ask for* a CT scan just because one wants to feel assured that one had a "thorough checkup." CT scans rarely produce important findings in people without relevant symptoms and there is a chance that the scan will find something incidental, spurring additional CT scans or X-rays that add to unnecessary dose.

Occupational Radiation Doses and Risks to Interventional Radiologists

Interventional radiologists receive non-uniform occupational radiation doses, with relatively high doses to the head and extremities and low doses to the trunk, which is protected by a lead apron. Twenty-eight interventional radiologists from 17 institutions wore thermo luminescent dosimeters (TLD) over their collars and under their aprons for a 2-month period. The estimated annual radiation dose was converted to effective dose as suggested by the International Commission on Radiological Protection. Effective dose is used to relate the risk associated with non-uniform dose to that associated with an equivalent uniform whole-body dose. The mean annual effective dose was 3.16 mSv (316 mrem), with a range of 0.37-10.1 mSv. The mean annual effective dose was approximately equal to the mean natural background dose of 3 mSv per year from radon and other natural sources and it was only 6% of the National Council on Radiation Protection and Measurements' recommended effective dose equivalent limit of 50 mSv per year. The annual radiation risk of fatal cancer would be less than one per 10,000 for almost the entire career of an interventional radiologist.

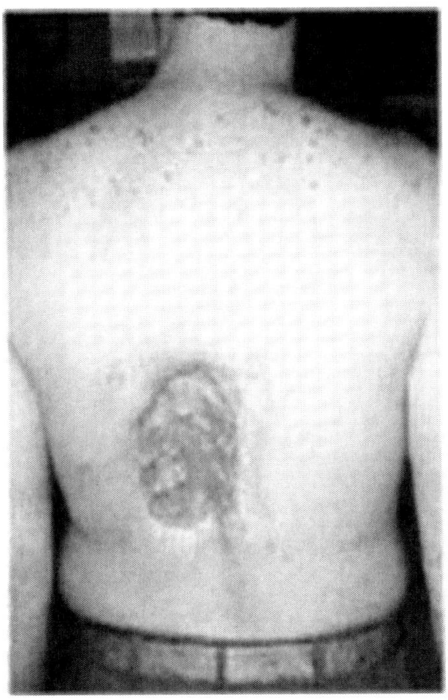

Figure. Appearance of radiation-induced skin injury approximately 18 to 21 months following multiple coronary angiography and angioplasty procedures - evidence of progressive tissue necrosis. (Source: www.fda.gov/cdrh/rsnaii.html.)

Table 1. Potential Clinical Effects of Radiation Exposures to the Skin and Eye Lens

	Effects	Threshold dose (Gy)	Time of onset
Skin	Early transient erythema	2	2-24 hours
	Main erythema reaction	6	~1.5 weeks
	Temporary epilation	3	~3 weeks
	Permanent epilation	7	~3 weeks
	Dermal necrosis	>12	>52 weeks
Eye	Lens opacity (detectable)	>1-2	>5 years
	Lens/cataract (debilitating)	>5	>5 years

Source: ICRP, 2000.

During a procedure, several major parameters influence dose:

- Number of images taken
- Fluoroscopy time, field size and overlap of fields (Miller 2002)
- Tube filtration, generator voltage and current
- Reduced-dose pulsed fluoroscopy versus continuous fluoroscopy (Wagner 2000)
- Distance between the X-ray tube and the patient and between the patient and the image receptor
- Patient body shape and size

Radiation dose is optimized when imaging is performed with the least amount of radiation required to provide adequate image quality and imaging guidance. Optimizing patient radiation dose also provides a direct benefit to the operator and assistants and scattered radiation dose in the room is directly proportional to the patient dose. If patient dose is reduced, so is the dose to the operator.

Radiation Risks from Interventional Fluoroscopy

The benefits of properly performed interventional fluoroscopy almost always outweigh the radiation risk experienced by an individual. However, unnecessary radiation exposure can produce avoidable risk to both the patient and the operator. The short-term risk to patients is radiation-induced skin damage, which can result from acute radiation doses of $\geq$ 2Gy. The extent of the skin injury may not be known for weeks after the procedure. Repeated procedures increase the risk of skin injury, because previous radiation exposure sensitizes the skin and cumulative higher doses lead to higher risk.

Long term effects include the potential risk of cancer. It is generally accepted that there is probably no low dose "threshold" for inducing cancers, i.e. no amount of radiation should be considered absolutely safe for this purpose Recent data from the atomic bomb survivors (Pierce 2000) and medically irradiated populations (UNSCEAR 2000) demonstrate small, but significant increases in cancer risk even at the level of doses that are relevant to interventional fluoroscopy procedures. The increased risk of cancer depends upon the age and sex of the patient receiving the radiation exposure. Children are considerably more sensitive to radiation than adults which has been consistently shown in epidemiologic studies of irradiated populations.

Risk to Health Care Providers

Health care providers are also at risk of radiation damage from chronic exposure to radiation from these procedures. There are an increasing number of case reports of skin changes on the hands and injuries to the lens of the eye in operators and assistants (Faulkner 2001). Although cancer is uncommon, cancers associated with radiation exposure in adults may include leukemia and breast cancer (Yoshinaga 2004).

Table 2. Strategies to Manage Radiation Dose to Patients and Operators

	Immediate	Long-Term
Optimize dose to patient	Use proper radiologic technique: Maximize distance between x-ray tube and patient Minimize distance between patient and image receptor Limit use of electronic magnification Control fluoroscopy time: Limit use to necessary evaluation of moving structures Employ last-image-hold to review findings Control images: Limit acquisition to essential diagnostic and documentation purposes Reduce dose: Reduce field size (collimate) and minimize field overlap Use pulsed fluoroscopy and low frame rate	Include medical physicist in decisions Machine selection and maintenance Incorporate dose-reduction technologies and dose-measurement devices in equipment Establish a facility quality improvement program that includes an appropriate x-ray equipment quality assurance program, overseen by a medical physicist, which includes equipment evaluation/inspection at appropriate intervals.
Minimize Dose to Operators and Staff	Keep hands out of the beam Use movable shields Maintain awareness of body position relative to the x-ray beam: Horizontal x-ray beam - operator and staff should stand on the side of the image receptor Vertical x-ray beam - the image receptor should be above the table Wear adequate protection Protective well-fitted lead apron Leaded glasses	Improve ergonomics of operators and staff: Train operators and staff in ergonomically good positioning when using fluoroscopy equipment; periodically assess their practice Identify and provide the ergonomically best personal protective gear for operators and staff Urge manufacturers to develop ergonomically improved personal protective gear Recommend research to improve ergonomics for personal protective gear

Strategies to Optimize Radiation Exposure from Interventional Fluoroscopy

An important goal of all interventional fluoroscopy is to achieve clinical success using the least amount of radiation consistent with adequate imaging guidance. However, most interventional procedures require high quality images, longer fluoroscopy time or both. Using appropriate operating parameters for x-ray machines will lower radiation doses to patients, and therefore to operators and assistants as well. It is critically important to adequately train operators and their assistants to use equipment that provides acceptable image quality along with the maximum possible dose-reduction, and to have equipment regularly inspected and maintained. Physicians, technologists, medical physicists, fluoroscopy equipment manufacturers and medical and governmental organizations share the responsibility to optimize radiation doses to patients undergoing interventional fluoroscopy.

Physician-Patient Communication for Interventional Fluoroscopy

Operators should always ask the patient about any previous history of interventional fluoroscopy before undertaking another procedure. It is important to communicate the details of the procedure, patient dose, and immediate and potential long-term health effects to patients and their primary care providers.

Before Procedure

Patients should be counselled on radiation-related risks, as appropriate, along with other risks and benefits associated with the procedure. If patients are likely to have multiple interventional fluoroscopy procedures in a short period of time, they should be informed if there is a possibility that significant radiation exposures may accompany these procedures and may cause potential short-term and long-term radiation-related health effects.

After Procedure

After a procedure, the measured or estimated radiation dose should be reviewed (Miller 2004), and appropriate steps should be taken to ensure adequate patient follow-up:

- A follow-up visit should be scheduled 30 days after the procedure for all patients who received a skin dose of 2 Gy or more or a cumulative dose of 3 Gy or more.
- Details of the interventional fluoroscopy procedure description, operative notes, doses and information about possible short-term and long-term effects should be sent to the patient's primary care provider.
- The patient and primary care physician should be specifically requested to notify the operator if observable skin effects do appear.

Dosimetry Records and Follow Up

Measure and record patient radiation dose: following detail should always be recorded:

- Fluoroscopy time
- Available measures like - DAP (Dose Area Product), cumulative dose, skin dose, etc.

It is the duty of operator to inform patients who have received high doses to self-examine the x-ray beam entrance site for skin erythema.

Mammography: This procedure may increase breast cancer risk in some women with some genetic mutations. Procedure may involve patients who may be in child bearing age or of relatively younger age group. It is this age group which may have higher probability of incidence of the breast cancer consequent to this exposure.

Mammography is an extremely useful non-invasive imaging technique used for the detection of breast cancer. It has played an important role in the screening of women above a certain age or with a family history of breast cancer. Nevertheless, a fundamental requirement of the International Basic Safety Standards for Protection against Ionizing Radiation (BSS) is the optimization of radiological protection of patients undergoing such medical exposure. Attempts to reduce doses to patients, however, must be balanced with the quality assurance considerations, to ensure X ray images to be of a good quality. It is emphasized by International Atomic Energy Agency (IAEA) that the instrument be of high standard and have the proper calibration and QA.

Study of mean glandular dose (MGD) during Diagnostic Mammography and the evaluation of the factors affecting it in Indian context were under taken. The data of the patients participating in the study has been collected on Mammomet 3000 Mammography unit of Siemen's make, which has auto exposure control and dose display mode. The dose data collected is based on various physical parameters of the patients such as:

- Weight and height, breast size
- Age group and ethnicity, etc.

It is very important to have pool of data on the above factors so that the dosage could be correlated with various contributory factors to ultimately minimized radiation dose exposure in such procedures without compromising the diagnostic quality. About 500 patients were included in this study comprising of following:

- Breast Cancer
- Fibro Adenoma of breast
- Fibrocystic Disease of breast
- Calcification in breast
- Duct Dilatation/Papilloma
- Galactocoel
- Haematoma in breast
- Fibroadenosis, etc.

The doses received by the patients in various conditions detailed above were measured and graphical representation of breast thickness vs. doses for different ages were obtained. These are the dosages as displayed by the inbuilt system of the mammography unit.

It was observed in our study that patients who underwent mammography with different abnormalities were receiving different doses, like breast cancer patients received doses in the range of 0.8 mGy - 3.8 mGy, Fibro adenoma 0.4 - 1.7 mGy, fibroadenosis and fibrocystic disease / breast cyst 0.3 – 1.6 mGy and benign breast diseases in the range of 0.4 – 1.0 mGy. Patients who were advised for FNAC / biopsy received doses in the range of 0.2 – 2.5 mGy and patients having no significant abnormalities received dose in the range of 0.3 - 1.0 mGy.

The doses received by the patients were found to be dependent on compressed breast thickness, For breast cancer patients 0.8 - 3.8 mGy for 7.9 cm to 3.8 mGy for 11.6 cm compressed breast thickness and for cases with fibrocystic disease a dose of 0.3 m Gy for 5.5 cm to 1.6 mGy for 6.2 cm compressed thickness of breast were observed.

After due examination it was found that 39.4 % (197/500) patients had no significant abnormalities, 11.2% (59/500) were suffering from fibro adenoma, 8.6 % (43 / 500) from fibroadenosis, 7.2 % (36 / 500) from fibrocystic disease / breast cyst and 6.8 % (34 / 500) were breast cancer patients in this study.

Revised Diagnostic Imaging Standards (July 2014) Thomas A. Mobilia and Melanie G. Gelfand, have reported the summary of new and revised diagnostic standards for accredited hospitals, critical access hospitals and ambulatory healthcare organizations that provide diagnostic imaging services.[11]

Medical experts have emphasized the importance of facilities utilizing diagnostic imaging to take necessary precautions to minimize the risk of radiation exposure. The fact that diagnostic imaging has become increasingly available to patients has also meant that the exposure to ionizing radiation has almost doubled in the past 20 years.[13] In the light of these statistics, there is a need to restructure radiation safety standards by focusing on areas such as CT protocols, development of incident management policies and procedures, and procurement of new technologies and software to track and monitor radiation dosage.[14] After recognizing the need to regulate quality and safety issues in radiology, this new set of rules will ensure better utility by focusing on patients safety, including radiation safety, oversight of imaging services, overall staff competency and equipment maintenance[16] with intended goal to "ensure that organizations providing imaging services have the requisite infrastructure and safety culture to minimize radiation exposure to patients and staff and provide safe and effective care."

The safety standards by US laws have addressed the requirements which include: (i) managing safety and security risks involving patients with special circumstances, such as claustrophobia or medical implants; (ii) risk management related to hazardous materials and waste, including assessment of staff radiation exposure levels by a radiation safety officer or medical physicist; (iii) annual performance evaluations of all imaging equipment by a diagnostic medical physicist; (iv) mandatory ongoing education for radiologic technologists, including training in radiation dose-reduction techniques; (v) minimum competency for radiology technicians, including registration and certification by July 1, 2015, by either The American Registry of Radiologic Technologists or the Nuclear Medicine Technology Certification Board; (vi) documentation of CT radiation doses in the patient's clinical record; and (vii) continuing collection of data on radiology incidents and injuries.[7]

These changes will be implemented in phases. The first phase, which had to take effect July 1, 2014, was focus on CT, nuclear medicine, PET and MRI. The second phase is to be implemented by July 1, 2015, and will focus on fluoroscopy, required qualifications for imaging clinicians, and cone beam CT used in dental offices and oral-maxillary surgery practices. While compliance with these new standards is important for all facilities seeking to maintain the Joint Commission accreditation of USA, it is equally important for the facilities and hospitals for their continued participation in Medicare and Medicaid reimbursement programs. These new developments only highlight the concern of scientific community for safety from low level diagnostic and imaging radiation doses.

CONCLUSION

Dose Estimation and Quality Control:

In medical practice the risk and benefit associated with the ionizing radiation are experienced by the same person. At individual levels these risks are usually small compared to the benefit with proper quality facility. However, when the total deterministic effect is projected for the population at large, the probability of the risk has a significant bearing on the total health structure. In order to reduce the doses to the patient's number of exposures need to be reduced.

While the benefits of computed tomography are well known in diagnosing diseases and trauma and in the guidance of interventional and therapeutic procedures, however benefits of these are not without risks.

Risk Evaluation:

A small amount of radiation energy when delivered to a biological system has a probability to increase the risk of cancer and adverse hereditary effects. It may produce and manifest in the life time of the individual exposed (somatic) or it may produce it effect in next generation (genetic). Some of the somatic effect which may be seen depend on the severity of radiation dose may be sterility (temporary or permanent), loss of appetite, impairment of vision (cataract), leukemia, blackening of skin, shortening of life span, carcinogenesis etc. To clarify this point further an analogy of infrared radiation (heat energy) with X-rays and gamma–rays would suffice for general perception. The heat contained in a sip of coffee (taken at 60° C and one sip of about 5-7 ml) if delivered in equivalent X-ray /gamma radiation it will be lethal (it will be equivalent to about 500-600 cGy).Therefore, the safe work practices (both for radiation worker and population) and containing the dose within threshold limit of harm or at minimal risk level has to be maintained.

Medical Auditing and Public Awareness:

The public awareness is very critical in radiological safety practices. There is a considerable scope for reduction of doses in diagnostic radiology. In pregnancy, special care needs to be taken. Once pregnancy is declared, the dose to the conceptuous should not exceed dose limit of 2 mSv during the remainder of the pregnancy. It is clearly understood that the low level of radiation exposure in Mammography, Cardiology, CT scan and contrast X-ray studies are large enough to warrant concern. Once the pregnancy is known X-ray exposure should be avoided to reduce the risk to the embryo and foetus. The estimation of organ doses and possible harm could be interpreted with the help of deterministic projections available in the cases being investigated. It is therefore essential to evolve a well calibrated response of the queries of patients and the risk involved in radiological investigations and the possibility of occurrence of the harm along with its severity be explained. Interventional fluoroscopy is an increasingly important and valuable tool for treating disease, but it is not without risk. It is important for the health care community, manufacturers and regulators to work together to optimize radiation doses to the patient and physicians must continuously think about optimizing radiation dose to the patient. Used prudently and optimally, interventional fluoroscopy is one of the valuable treatment modalities for a wide variety of diseases and disorders if the above concern about the safety and risk are addressed adequately.

References

[1] Faulkner,K. and Vaño,E. Deterministic Effects in Interventional Radiology. *Rad. Prot. Dosim* 2001; 94:95-8.

[2] International Commission on Radiological Protection. Avoidance of radiation injuries from medical interventional procedures. *ICRP Publication* No. 85. Ann ICRP 2000;30:7-67.

[3] Koenig TR, Wolff D, Mettler FA, Wagner LK. Skin injuries from fluoroscopically guided procedures. *Am. J. Roentgenol.* 2001; 177:3- 20.

[4] Mettler F, Koenig TR, Wagner LK, Kelsey CA. Radiation injuries after fluoroscopic procedures. *Seminars Ultrasound*, CT, MRI 2002; 23:428-42.

[5] Miller DL, Balter S, Noonan PT, Georgia JD. Minimizing radiation-induced skin injury in interventional radiology procedures. *Radiology* 2002; 225:329-36.

[6] Miller DL, Balter S, Wagner LK, et al. Quality improvement guidelines for recording patient radiation dose in the medical record. *J. Vasc. Interv. Radiol.* 2004; 15:423-9.

[7] Pierce DA, Preston DL. Radiation-related cancer risks at low doses among atomic bomb survivors. *Radiat Res.* 2000; 154:78-86.

[8] Sources and Effects of Ionizing Radiation. United Nations Scientific Committee on the Effects of Atomic Radiation, UNSCEAR 2000 Report to the General Assembly, with Scientific Annexes, Volume II: Effects. New York: United Nations, 2000.

[9] Wagner LK, Archer BR, Cohen J. Management of patient skin dose in fluoroscopically guided interventional procedures. *J. Vasc. Interv. Radiol.* 2000; 11:23-33.

[10] Yoshinaga S, Mabuchi K, Sigurdson AJ, Doody MM, Ron E. Cancer risks among radiologists and radiologic technologists: Review of epidemiologic studies. *Radiology* 2004; 233:313-21.

[11] http://www.jointcommission.org.

[12] *Radiation Risks of Diagnostic Imaging*, 47 The Joint Commission Sentinel Event Alert 1, (2011) (hereinafter "Sentinel Event Alert").

[13] Neomi Mullens, *Are You Ready to Comply with New Radiation Safety Rules?*, AuntMinnie.com (February 6, 2014);http://www.auntminnie.com.

[14] Elizabeth Eaken Zhani, *Joint Commission Announces New and Revised Diagnostic Imaging Standards*, The Joint Commission (February 6, 2014); http://www.jointcommission.org Zhani, *supra* note 6.

[15] See The Joint Commission Prepublication Standards at http://www.jointcommission.org.

[16] *Medicare Coverage of Imaging Services*, Department of Health and Human Services: Centers for Medicare & Medicaid Services, June 2013, available at http://www.cms.gov.

In: Biological Responses, Monitoring and Protection ... ISBN: 978-1-63463-852-4
Editor: K. P. Mishra

Chapter 6

EPIDEMIOLOGY AND MOLECULAR STUDIES IN CEREBROVASCULAR DISEASE AT THE LATE PERIOD AFTER RADIATION EXPOSURE IN CHERNOBYL

D. Bazyka, V. Buzunov, I. Ilyenko and K. Loganovsky
National Research Center for Radiation Medicine (RCRM),
Kyiv, Ukraine

ABSTRACT

Consensus on the radiation effects in cardiovascular system doesn't exist. More evidence is needed to conclude whether or not radiation exposure has increased the incidence of cardiovascular and cerebrovascular disease, and associated mortality. The observed excess of cerebrovascular diseases is linked to those having worked during the first six months and having cumulated doses of more than 150 mSv. Although these results are statistically compatible with those of the atomic bombings survivors study, they could only be regarded as preliminary. Latest hibakusha data supposes an increase in mortality. Industrial cohorts' studies have controversial results. According to UNSCEAR, "There are deterministic effects, such as cataracts or cardiovascular disease, for which radiation exposure is not the only known cause. If these effects occur, usually some time after high levels of exposure, and there is no specific marker for radiation exposure having caused them, it is not possible to attribute the effect with certainty to radiation exposure, but only to express a probability that radiation was wholly or partly the cause,"[1].

The existing level of knowledge allows considering the contributive role of the immune function changes of the late period after irradiation to the cerebrovascular pathology pathways. Therefore, it is also actually taking into account an interaction between the immune and nervous systems. Encephalopathy ("radiation encephalopathy", "discirculatory encephalopathy"- DEP, "demielinisation" encephalopathy) according to medical statistics is one of the most common disorders of a late period after irradiation and one of the most frequent diagnosis for establishing a causal connection of disease with a radiation factor. At the same time, the nature of this disorder is not identified, and many researchers do not recognize it as an effect of irradiation. In some studies it was depicted as a result of direct brain tissue radiation damage, or as an indirect effect due to damage of vascular circulation. Development of the brain pathology as a result of CVD and vegetative dysfunctions, and the substantial vasculatory changes revealed in encephalopathy patients provide support to a latter assumption. More evidence is needed to conclude whether or not radiation exposure has increased the incidence of cardiovascular and cerebrovascular disease and associated mortality [1].

Epidemiology

Recent meta-analysis of data obtained in several countries has supported an association between circulatory disease mortality and low and moderate doses of ionizing radiation. However analysis was limited by heterogeneity among studies (particularly for non-cardiac endpoints), the possibility of uncontrolled confounding in some occupational groups by lifestyle factors, and higher dose groups (>0.5 Sv) generally driving the observed trends [2].

RCRM data have demonstrated increased radiation risks of incident cerebrovascular disease, including cerebral atherosclerosis and hypertensive encephalopathy. Ionizing radiation is only one of the set of factors inducing CVD.

In post-accident years, we carried out numerous investigations to assess the health effects of the Chernobyl accident. Epidemiological studies have found a significant increase in a wide range of non-tumor diseases [3, 4]. Among these circulatory diseases prevail, now the last mentioned diseases approximate to 30 % in the total incidence structure and 70 % in the total mortality structure.

In view of the high prevalence of circulatory diseases among the Chernobyl clean-up workers exposed to low doses of short-term ionizing radiation (up to 1 Gy), our studies of the development of the pathology are extremely important both theoretically and practically.

Incidence study was performed in clean-up workers cohort exposed in 1986-1987 for the period 1988-2010. Data from the State Registry of Ukraine (SRU) of Persons Affected by the Chernobyl Accident were used as the main source of information. SRU database contains 314,192 clean-up workers (216,031 of them are participants of recovery operations in 1986-1987). Study cohort included male clean-up workers 1986-1987 having information on whole-body external doses (68,145 people).

Table 1 presents the total number of examined clean-up workers 1986-1987 depending on age at the time of exposure (aged <40 and ≥ 40) and whole-body external doses by five-year observation periods.

The cohort was subdivided into:

- two age groups (<40 and ≥ 40 years at the time of exposure);
- dose subgroups (0-0.049; 0.5-0.099; 0.1-0.199; 0.2-0.240; 0.25-0.69 Gy).

Epidemiological study covered the observation periods (1988–1992, 1993–1997, 1998–2002, 2003–2007, 2008–2010). In the course of descriptive analysis, we calculated the incidence rate and 95% confidence interval. To assess the influence of radiation risk factor, absolute (ARY) and relative (RRY) risks as well as excess relative risk (ERR) were calculated. Dose group 0–0.049 Gy (internal control) was used as a control.

Table 1. The number of examined persons in dose-age subgroups of the cohort (clean-up workers 1986-1987) for 5-year observation periods

Period (years)	Age at the time of exposure									
	18-39					40-60				
	whole-body external dose (Gy)					whole-body external dose (Gy)				
	0-0.049	0.05-0.09	0.1-0.199	0.2-0.249	0.25-0.7	0-0.049	0.05-0.09	0.1-0.199	0.2-0.249	0.25-0.7
cohort	6233	17798	14708	14625	6460	925	1687	1395	1085	708
1988-1992	15370	50387	45861	48138	19744	2656	6331	5157	4339	2264
1993-1997	23198	67428	55659	56485	24532	4105	8845	6569	5433	3366
1998-2002	21434	63411	53459	54175	23688	3543	8046	6016	5021	3043
2003-2007	20142	47819	52460	50310	22156	2887	6794	4975	4145	2521
2008-2010	10335	30963	26951	27732	12305	1179	3040	2290	1942	1131
Total	90479	260008	234390	236840	102425	14370	33056	25007	20880	12325

Figure 1 presents the dynamics of circulatory diseases incidence in clean-up workers 1986-1987 for 24-post-Chernobyl years. The highest rates of those were found over 12-16 years after the exposure. Since 2003, a certain fall of rates has been registered (mainly among the people aged 40 and older at the time of exposure). The latter can be explained by a decrease in the cohort size (due to increased mortality) and also by earlier diagnostic. Here in younger patients, incidence at the final stages of observation is rather high, demanding for more precise diagnostic and therapeutic procedures.

The following – chronic ischemic heart disease, cerebrovascular diseases, hypertension, diseases of blood vessels (arteries, arterioles, capillaries and lymph vessels) prevail in the structure.

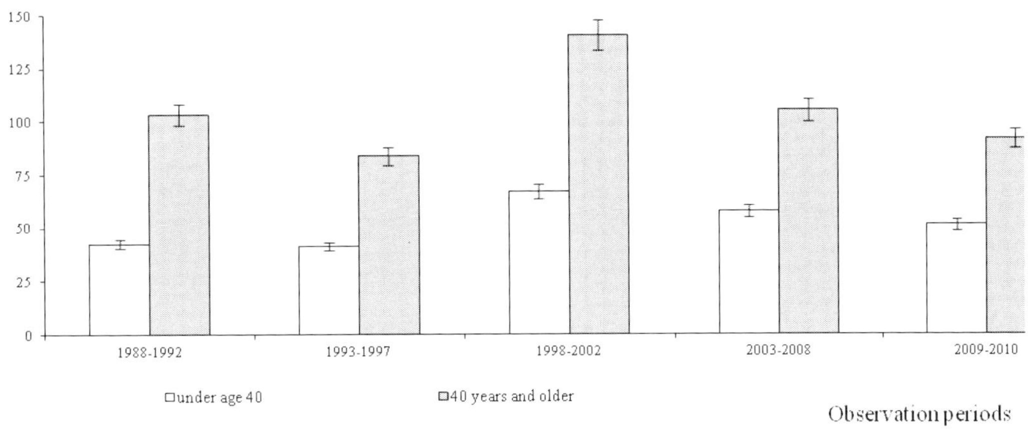

Figure 1. The dynamics of circulatory diseases incidence (ICD codes I00.0-I99.9) in clean-up workers 1986-1987 depending on age at the time of exposure.

To determine the effect of radiation and non-radiation factors on the development of circulatory diseases we carried out the epidemiological analysis and risk assessment. In clean-up workers 1986-1987, there were established significant relative risks for the diseases (ischemic heart disease, cerebrovascular diseases, diseases of arteries, arterioles and capillaries, diseases of veins, lymphatic vessels and lymph nodes) in the range of whole-body external doses 0.05-1.0 Гр. The stable risk values were obtained for clean-up workers 1986-1987 under age 40 at the time of exposure (doses 0.05-0.7 Gy) and in the elder age group 0.25-0.7 Gy (Table 2). The findings prove higher sensitivity of younger people to ionizing radiation. However, this issue requires further investigation.

It is noted that the risk of circulatory diseases is largely due to the influence of non-radiation factors, among these are adverse working conditions before the Chernobyl accident, emotional stress, physical overstrain, smoking, alcohol abuse, malnutrition and heredity (Table 3).

Among clean-up workers, the age factor in circulatory diseases is predominant. Thus, for clean-up workers aged 40-49 at the time of examination, the cerebrovascular incidence significantly exceeds that for people of 30-39 years old (2.4 times), for those aged 50-59 (5.2 times) and for people of 60-69 years (7.7 times). The significant relative risks are: RR = 2.4; 5.2; 7.7. In these age groups, the relative risks for ischemic heart disease equal 4.6; 7.4; 7.9, for diseases of arteries and arterioles – 6.0; 14.3; 16.1 respectively (calculations were based on CER data).

Genetic predisposition has been identified in 31.7%, body mass excess had been found in 38%, smoking in 55.2%. Mental health studies have shown elevated rates of cognitive disorders in exposed including depression, suicide ideation and high alcohol abuse rates both in exposed and non-exposed. The observed excess of cerebrovascular diseases after Chornobyl is linked to those having worked during the first six months and having cumulated doses of more than 150 mSv [2].

Table 2. Risks for circulatory diseases in clean-up workers 1986-1987 depending on whole-body external dose and age at the time of exposure over 24-year post-accident period (RRY with 90 % CI is in italics)

	under age 40 at the time of exposure		40 – 60 years at the time of exposure	
Dose (Gy)	RRY (95 % CI)	ERR (Gy^{-1})	RRY (95 % CI)	ERR (Gy^{-1})
Circulatory diseases (ICD-10 codes I00.0-I99.9)				
0.05 – 0.099	*1.35 (1.32; 1.37)*		1 (0.91; 1.1)	
0.1 – 0.199	*1.19 (1.11; 1.27)*		*1.12 (1.1; 1.18)*	
0.2 – 0.249	*1.18 (1.16; 1.2)*		*1.,07 (0.96; 1.2)*	
0.25 – 0.7	*1.37 (1.14; 1.34)*	0.8 (0.57; 1.0)	*1.15 (1.02; 1.3)*	0.3 (0.17; 0.43)
Ischemic jeart disease (ICD-10 codes I20.0-I25.0)				
0.05 – 0.099	*1.38 (1.33; 1.42)*		0.87 (0.75; 1.0)	
0.1 – 0.199	*1.18 (1.02; 1.36)*		0.99 (0.92; 1.1)	
0.2 – 0.249	*1.27 (1.23; 1.32)*		0.95 (0.8; 1.13)	
0.25 – 0.7	*1.32 (1.27; 1.37)*	0.7 (0.52; 0.97)	0.96 (0.8; 1.16)	
Cerebrovascular diseases (ICD-10 codes I60.0-I69.8)				
0.05 – 0.099	*1.27 (1.22; 1.33)*		0.92 (0.76;1.1)	
0.1 – 0.199	*1.01 (0.83; 1.22)*		*1.06 (0.95; 1.2)*	
0.2 – 0.249	*1.07 (1.02; 1.12)*		*1.07 (0.86; 1.3)*	
0.25 – 0.7	*1.51 (1.44; 1.59)*	1.1 (1.0; 1.3)	*1.18 (0.9; 1.48)*	
Diseases of arteries, arterioles and capillaries (ICD-10 codes I70.0-I79.8)				
0.05 – 0.099	*1.48 (1.38; 1.59)*		0.97 (0.7; 1.32)	
0.1 – 0.199	*1.22 (0.9; 1.65)*		*1.1 (0.9; 1.32)*	

	under age 40 at the time of exposure		40 – 60 years at the time of exposure	
Dose (Gy)	RRY (95 % CI)	ERR (Gy^{-1})	RRY (95 % CI)	ERR (Gy^{-1})
0.2 – 0.249	*1.14 (1.07; 1.23)*		0.92 (0.6; 1.32)	
0.25 – 0.7	*1.65 (1.52; 1.78)*	1.4 (1.12; 1.7)	*1.44 (1.01; 2.06)*	0.9 (0.71; 1.12)
Diseases of veins, lymphatic vessels (I80.0-I89.9 за МКХ-10)				
0.05 – 0.099	*1.39 (1.32; 1.46)*		*1.24 (0.93; 1.64)*	
0.1 – 0.199	*1.2 (0.97; 1.49)*		*1.3 (0.9; 1.72)*	
0.2 – 0.249	*1.18 (1.12; 1.24)*		*1.25 (0.9; 1.72)*	
0.25 – 0.7	*1.4 (1.32; 1.48)*	0.8 (0.62; 0.97)	*1.45 (1.03; 2.02)*	0.9 (0.79; 1.1)

The combined effect of radiation and non-radiation factors may significantly exceed that of radiation.

Table 3. Significant (95 % CI) relative risks for certain nosological forms of circulatory diseases among clean-up workers 1986-1987 (due to several non-radiation factors)

Name of the disease	Adverse working conditions before the Chernobyl accident	Emotional stress	Smoking	Alcohol abuse	Physical overstrain	Malnutrition
hypertension (401-405)	2.3 (1.3; 4.4)	2.1 (1.5; 3.1)	1.6 (1.2; 2.3)	1.9 (1.5; 2.5)	1.4 (1.2; 1.5)	2.9 (2.6; 3.2)
IHD (410-414)	2.4 (1.8; 6.1)	4.5 (2.5; 5.5)	2.6 (1.6; 4.6)		1,7 (1.1; 2.7)	
cardiomyopathy (425)	1.7 (1.1; 2.7)	1.6 (1.2; 2.8)		1.3 (1.2; 1.4)	1.4 (1.2; 1.6)	1.6 (1.3; 2.0)
cerebrovascular disease (430-438)	2.3 (1.8; 3.0)	2.4 (1.6; 3.1)	1.9 (1.5; 2.3)		1.6 (1.4; 1.9)	1.5 (1.3; 1.9)

Thus, our results suggest the following:

In clean-up workers, the post-accident period is characterized by significant growth of circulatory diseases, especially cerebrovascular diseases (due to the cases of stroke, cerebral infarction, cerebral atherosclerosis, hypertensive encephalopathy), ischemic heart disease (due to increase in chronic ischemic heart disease, acute myocardial infarction, angina pectoris), hypertension, diseases of arteries, arterioles and capillaries (owing to the cases of atherosclerosis, embolism, arterial thrombosis), diseases of the veins and lymphatic vessels (phlebitis and thrombophlebitis, venous thrombosis and embolism).

Higher incidence rates observed 12-16 years after the exposure suggest of a latent period in the development of the pathology.

The results of risk-analysis revealed certain dose-dependent effects. We obtained significant values of relative risks in the range of whole-body external doses 0.05-0.7 Gy. It should be noted that the most stable values of risks and excess relative risks were obtained at doses 0.25-0.7 Gy.

Younger people exhibited higher sensitivity to ionizing radiation. However, issue of age radiosensitivity requires further investigation.

Our data agree to a certain extent with the results of health monitoring of hibakushi – Hiroshima and Nagasaki A-bomb survivors [5-6]; significant relative risks for some circulatory diseases in clean-up workers were also proved by studies of Russian scientists [7]. So far we have collected enough data which is evident of the effect of low-dose radiation (both external and internal) on the development of non-tumor incidence (primarily cardiovascular and cerebrovascular), still there is no consensus on the problem. Therefore, further study of the problem is considered to be relevant and important.

Molecular Studies

Ionizing radiation is one of factors of modification of the lymphocyte function with oxidative damage, DNA breaks, and resulting changes of proliferation, apoptosis and cellular senescence. Telomeres are demonstrated to be a critical element of cell arrest, proliferation, aging and malignant transformation. Aim of molecular studies was to reveal markers that could be attributable to ionizing radiation dose at the late period after exposure and contribute to the cerebrovascular pathology (CVD) and cognitive dysfunctions.

Study of dose dependencies of the immune system parameters was conducted in CVD patients with available dose estimates. We used official data on doses from the State Chornobyl registry data and doses from Chornobyl NPP dosimetry shop. Analysis of immunological parameters depending on the time of irradiation showed that at acute exposure and time after exposure greater than 10 years, the statistically significant differences were determined depending on the dose, not the time. Doses, the effect of which could be detected at this period were 0.25 – 0.35 Gy. However, for chronic exposure of staff of a 30-km exclusion zone, the major factor was the duration of work activity in the zone, and the threshold dose was reduced. Examination of clean-up workers with CVD has demonstrated in 42.3% of cases the decreased CD3+ and CD4+ cell counts in peripheral blood. Depression of cellular immunity was association with the preservation of the increased variant TCR-cell counts.

The critical function of telomeres consists in adjusting the cellular response to stress and growth stimulation based on previous cell divisions and is accomplished by progressive telomere shortening and DNA damage responses activated by chromosome ends without sufficient telomere repeats [8]. Ionizing radiation in low-dose is regarded as one of factors modifying the cellular systems with oxidative stress, DNA damage, cellular senescence, apoptosis and compensatory proliferation which cause the change life-span [9]. These factors are believed to represent genome instability and be a background for radiation aging. Also telomere shortening may serve as a surrogate marker of the progression of malignant disorders and seems to be accelerated in allogenic bone marrow transplant recipients [10]. In radiation biology this question is not clearly understood. Neuhof et al. suggest telomerase activity to be the radiation modified function, responsible for the damage DNA. Length of telomeres is bound also to enhanced chromosome radiosensitivity that is demonstrated in C3H mice acute myeloid leukemia cell line 7926 with a doubled telomere length of telomeres and in comparison with 8709 cell line [11, 12]. In radiosensitive cells short telomeres might correlate with radiation sensitivity [13]. Such conclusion could be applied to such radiosensitive cells as human lymphocytes and other types of leukocytes. Some other data show that long telomeres, but not telomerase activity seem be factor protecting proliferation activity under radiation exposure [12, 14].

Presence of radiation senescence in people who suffered as a result of Chornobyl accident, behaves to the most debatable aspects of biological effects of irradiation. Individuals who suffered as a result of the Chornobyl accident might have genome changes, changes of linear and functional active cellular biomolecules, such as lack of CD28 on CD8+ cells, which are experimentally shown to result in lower proliferative capacity, decreased IL-2 production, telomere erosion and less response to TCR stimulation [15]. It is actual to reveal the mechanisms which can result in the radiation induced senescence, or possibly malignant transformation onset, and a direct role of losses of length of telomeres and apoptosis. Due to the emerging data on such questionable radiation effects in humans as cell senescence, of importance is to obtain comparable results using the quantitative approach.

Molecular studies were performed in 136 cleanup workers randomly selected from the Chornobyl State registry (table 4). Control group included 22 healthy individuals of the same age (53,1±1,4 years) with haematological and immunological indices within the age limits. All patients have verified doses of exposure and were divided in for groups according to the dose limits: I: 0-20 mSv, II – 20 – 100 mSv, III – 100 – 250 mSv, IV – 250- 500 mSv.

Table 4. Cleanup workers distribution by age and dose of exposure

Study groups	nn	Age, years	Dose, mSv
		M±m	M±m
I	21	63,2±1,9	9,3±1,2
II	43	58,9±1,0	53,0±2,7
III	45	59,0±1,2	184,3±7,0
IV	27	62,7±1,3	418,0±44,1

All subjects have participated at the study by the informed consent.

Study of radiation induced changes of T-cell receptor has shown no tendency of changes after exposure to doses less than 350 mSv.

Study of the RTL was conducted in peripheral blood lymphocytes by the flow-FISH assay (combining the fluorescent hybridization *in situ* and flow cytometry) with a Telomere PNA Kit/FITC (Dako Cytomation, Denmark). This kit is recommended for the detection of telomeres in nucleated haematopoietic cells using a FITC–conjugated peptide nucleic acid (PNA) probe. Results are evaluated by flow cytometry using a light source with excitation at 488 nm.

On a single cell suspension (consisting of a mixture of sample cells (peripheral blood lymphocytes) and control cells (K562 cell line – Bank of cell lines from human and animal tissue, R.E. Kavetsky Institute of Experimental Pathology, Oncology and Radiobiology of Nat. Acad. Sci. of Ukraine, Dept. Experim. Cell Systems), the sample DNA is denatured for 10 minutes at 82 °C in a microcentrifuge tube either in the presence of hybridization solution without probe or in hybridization solution containing fluorescein-conjugated PNA telomere probe. Then hybridization takes place in the dark at room temperature overnight. The hybridization is followed by two 10-minute post-hybridization washes with a Wash Solution at 40 °C. The sample is then resuspended in an appropriate buffer for further flow cytometric analysis. Samples hybridized with the Telomere PNA Probe/FITC should exhibit a fluorescence signal in FL1, which is higher than the background/autofluorescence signal obtained from the sample of the same cells hybridized with the hybridization solution without probe. When analyzing flow cytometry data for cells hybridized with or without the Telomere PNA Probe/FITC it is important only to look at cells in the G0/1-phase of the cell cycle where the cell has one copy of its genome. The gated cells can then be displayed in the FL3-height versus FL1-height dot plot. After flow cytometric analysis, the data obtained can be used for determination of a relative telomere length (RTL). The RTL value is calculated as the ratio between the telomere signal of each sample and the control cell (K562 cell line) with correction for the DNA index of G0/1 cells. The RTL value was calculated by a formula (1):

Formula 1.

$$RTL = \frac{(\text{mean } FL1_{exp.cells} \text{ with PNA}) - (\text{mean } FL1_{exp.\ cells} \text{ without PNA})}{(\text{mean } FL1_{ctrl\ cells} \text{ with PNA}) - (\text{mean } FL1_{ctrl\ cells} \text{ without PNA})} \times 100$$

The differential analysis of apoptosis was conducted by the Annexin V-FITC Apoptosis Detection Kit I (BD, USA) by flow cytometry. General Annexin V Staining Procedure: wash cells twice with cold PBS and then resuspend cells in 1X Binding Buffer (0,1 M Hepes/NaOH (pH 7,4) 1,4 M NaCl, 25 mM $CaCl_2$) at a concentration of ~1 x 10^6 cells/ml. Transfer 100 μl of the solution (~1 x 10^5 cells) to a falcon tube. Add 5μl Annexin V and 5μl PI. Gently mix the cells and incubate for 15 min at room temperature in the dark. Add 400 μl of 1X Binding Buffer to each tube. Analyze by flow cytometry as soon as possible (within 1 hr).

Analysis carried out on a FACSCalibur (BD, USA) flow cytometer. Batch analysis of flow cytometry data was performed in Lysis II and WINMDI 2.8 software, statistical analysis of the results was conducted using SPSS 16.0. package.

Performed study has shown a decrease of the mean RTL parameters in all groups of radiation exposed. However, the substantial decrease was typical for a group of Chornobyl cleanup workers. There was a substantial tendency of telomere shortening with age (Figure 2). The longest telomeres were associated with the lower percentage of cells with variant T-cell receptors (Figure 3). Figure 1 demonstrated a decrease in RTL of PB leukocytes depending on the research group.

Earlier telomeres in a group of 181 individuals were associated with in vitro radiosensitivity as the group with short telomeres presented higher frequencies of ionizing radiation-induced micronuclei when compared to the long telomeres group and the individual telomere length is regarded as biomarker of individual chromosome instability upon exposure to ionizing radiation [16]. Later study hasn't shown differences cell cycle regulatory proteins (cyclins A, B1, D1, E, and Ki67) and telomere length in radiation-induced and sporadic papillary thyroid carcinoma. Coming from the theory of cellular senescence, which is bound to reduction of telomere ends of chromosomes and our findings, the two ways of realization of cellular cycle could be probable: cells can be eliminated by apoptosis, or to move to a "senile" phenotype. The last type could be of some positive importance, as a cellular senescence dereasing the proliferation rates and thus blocking cancerogenesis. Vice versa, an accumulation of genetic errors at senescence could also lead to cells transformation.

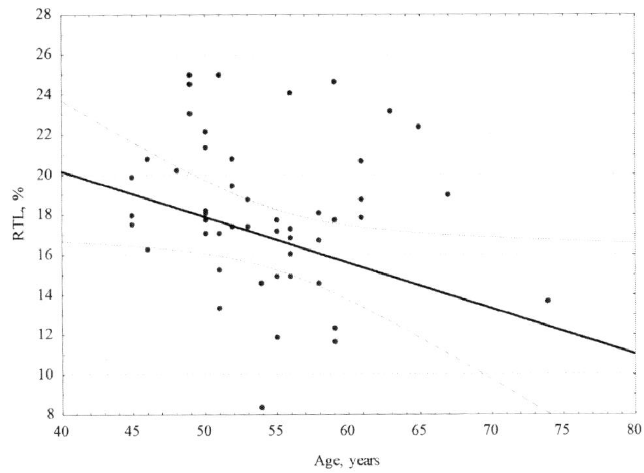

Figure 2. Dependency between the relative telomere length and age.

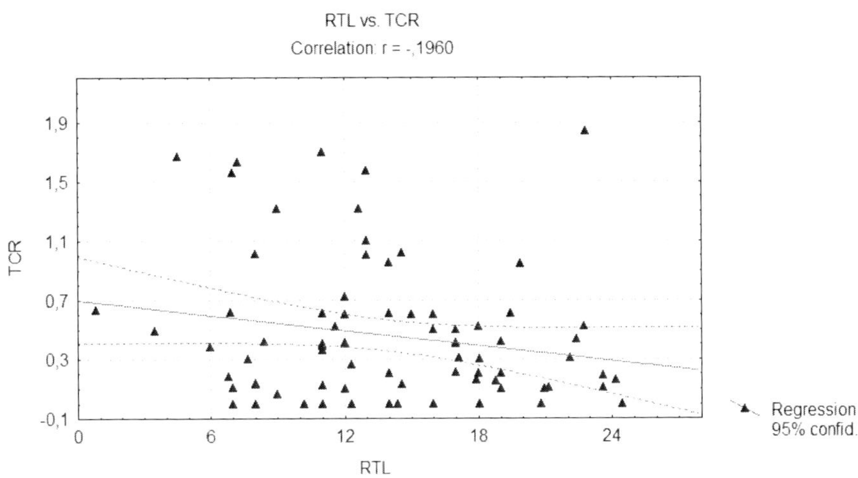

Figure 3. Dependency between the relative telomere length and variant TCR cells counts.

The other way is a programmed cellular death. Realization of apoptosis includes a cascade of reactions, in which one of main roles is played by TP53 gene and it product - p53. However, in cells with p53 mutations, genome instability is increased, which promotes malignant transformation. Malignant cells overexpress telomerase that lengthens telomeres and allows for plentiful divisions. Telomere elongation is a common molecular feature of advanced malignancies, but there is some evidence suggesting that short telomeres and concurrent chromosomal instability contribute to malignant cell transformation [17].

In a prospective, population-based study on 787 participants for a 10-year period to estimate the impact of telomere length on overall cancer manifestation and mortality it has been demonstrated that there is a significant inverse correlations between baseline leukocyte telomere length and both cancer incidence and mortality, which emerged as independent of standard cancer risk factors. There was evidence of heterogeneous effects according to cancer type, whereby tumors with a high fatality rate: gastric, lung, and ovarian cancer tended to exhibit more prominent relationships with telomere length and tumors with a more favorable prognosis showed modest or no associations [18].

A tendency of the mean telomere length parameters was demonstrated in all groups of studied cleanup workers. Significant changes were shown in exposed to in dose interval between 20 and 250 mSv (table 5). A question appear, is it possible to consider such process as an increased replication senescence, or radiation-induced cellular senescence? In replication senescence the somatic cells don't enter division and in vitro normal somatic cells are aging after several mitoses. Such cells are blocked in G1-phase, however not capable for replication; they remain metabolically active, viable and are radiation resistant. hTERT activity results in telomere lengthening, thus protecting cell from replication aging. Cells passing through the G1- checkpoint acquire radiation hypersensitivity. In addition to replication aging, cells undergo senescence. Such mechanism works in vitro, and shows a dose-dependent decline of RTL index, so similar to our results in vivo and possibly reflecting a stress-induced phenomenon. Such changes could be associated with apoptosis or other systemic induced cellular reactions. Our data show that there are marked changes in apoptosis, its early and late stages. The highest numbers of annexinV+PI- cells in early apoptosis and annexinV+PI+ cells in late apoptosis have been detected in a remote period after irradiation in a group of Chornobyl clean-up workers.

Table 5. Relative telomere length in Chornobyl cleanup workers

Groups	M	±SD	Student's P
1. Cleanup workers, <20 mSv	17,26	4,82	0,2
2. Cleanup workers, 20-100 mSv	16,31	5,02	0,059
3 Cleanup workers, 100- 250 mSv	15,87	4,39	0,005
4. Cleanup workers, 250-500 mSv	16,96	4,67	0,13
Control	18,94	5,73	

Relationship was demonstrated between the RTL and apoptosis. The groups of persons with higher doses are characterized by the declined RTL. For a annexin+PI- fraction, however these changes are of an opposite type. The lower RTL values are associated with the higher apoptosis levels and of apoptosis and to a maximal extent such dependency is realized in a group of clean-up workers. Such type of response could be explained, as such factor, as reduction of telomere length together with oxidative cell damage and genome instability might result in elimination of defective cell by apoptosis. There are suggestions that telomere shortening might be the main or even the sole mechanism of the natural and radiation aging.

Latest studies show that different factors modify telomere length as well as telomerase activity and, probably, biological age also. Most of somatic progenitor cells show progressive telomere shortening with each round of replication; however B-lymphocytes appear to be an exception to this rule in that memory B cells have longer telomeres than naive B cells. Elongation of telomeres in the B-cell lineage could reflect a requirement for extensive cell divisions imposed by clonal selection and affinity maturation of antibodies. Could it be that B cells are more likely to form tumors than T cells as a result. In general, the difference in telomere biology between the cells of the germ line and various somatic (stem) cells is poorly understood. Linear correlation analysis shows in cleanup workers correlation between telomere length and size of CD34+ early progenitor cel subset but not with the numbers of mature cells (CD3+, CD19+).

Our study shows that there is a substantial change in telomere length as in the early and also in a late period after radiation exposure and these changes are related to changes of apoptosis. Data of our and other studies at the early period after radiation exposure demonstrate that ionising radiation induces a complex response including not only DNA, but a set of epigenetic changes, oxidative damage, membrane viscosity and expression of surface antigens molecules accompanied by somatic mutations in exposed cells.

The cognitive disorders in clean-up workers of the Chornobyl accident gain more clinical and social significance. The definition of the diagnostically significant criteria of the cognitive functions impairment in cerebrovacular pathology and neurodegenerative brain disorders as well as search for specific immunological markers which may be a basis for individual health care and prophylactic programs for this group of suffered after the Chernobyl accident became particularly relevant [19].

Numerous experimental and clinical studies have shown that ionizing radiation might be one of the cognitive impairment triggers, which severity depends on its intensity. Formation of these effects in remote post-radiation period might be a sign of stable molecular disturbances initiated during contact with ionizing radiation, that continue to change the native coordination of systems, including the nervous, endocrine, immune ones, which synergism provides protection and cognitive field function in the irradiated organism. This assumption is confirmed by the recent years facts of cognitive impairment increased on a background of the organic brain injury in clean-up workers of the Chernobyl accident. Of major importance are the genetic instability and transcription changes, immune deficiency and activation of pro-inflammatory and proapoptotic mechanisms.

Hereby, the study of relationships between immunological parameters changes and cognitive impairment in estimation of radiation effects in the clean-up workers at the remote period after irradiation gains significance.

The study in clean-up workers on a ground on standard neurological examinations and psychiatric interview using the following psychiatric scales: Brief Psychiatric Rating Scale (BPRS) was used for unify quantitative assessment of psyhoneurologic symptoms and psyhopathology in general; Mini-mental state examination (MMSE) was used for cognitive level screening diagnostic.

For qualitative and quantitative assessment of structure of psyhopathologic syndrome, specific psychic function and personality and intellect disorders the following was used: the General Health Questionnaire (GHQ-28) - for study of somatoform symptoms, anxiety/ insomnia, social dysfunction and depression severity; the Zung Self-Rating Depression Scale (SDS) - for determination of unmasked depression; the Impact of Events Scale (IES) - for evaluation of posttraumatic stress disorders (PTSD) and for excitement evaluation in connection with PTDE - clinical scale for irritation self-assessment - Irritability, Depression, Anxiety (IDA).

In assessing the mental status it was revealed that 86.7 % of the clean-up workers exposed to >100 mSv had signs of organic mental disorders, compared with 66.7 % of all examinees in the group with a dose range 20-100 mSv. The MMSE screening test revealed no statistically significant difference in the study groups. At the same time, the average score in all four groups was lower against normal value that could be an indicator of the cognitive impairment presence in most patients. While the addition psychometric tests (BPRS, IES,

IDA, GHQ- 28, SDS) it was determined that with the increasing of radiation dose in examinee groups against the group expose to <100 mSv the intensity of psychopathological disorders, particularly higher levels of anxiety, depression, and also indicators of somatoform disorders, anxiety/insomnia and social dysfunction was increased [20].

While assessing the changes in immunologic parameters it was revealed that percentage of CD3+HLA-DR+ cells and TcRαβ+/γδ+ cells ratio were higher in exposed; CD25+ Tregs counts25 inversely correlated with radiation dose ($rs = -0.615$; $p = 0.033$) in groups of the clean-up workers exposed to >100 mSv. CD95+ cell counts, Fas-mediated apoptosis marker, were significantly higher in the patients with radiation doses >250 mSv both comparing with controls and groups of examinees with lower radiation doses. There is a weak, but statistically significant correlation between cognitive function (MMSE) and relative telomere length [20]. Better cognitive abilities are associated with longer telomeres, or by other words, telomere shortening is associated with cognitive deficit (table 6).

Table 6. Relative telomere length in the groups of Chornobyl cleanup workers with cerebrovascular pathology associated with cognitive deficit

Groups		RTL (%) M±SD	p
Non-exposed	Control	18,94±5,73	
	CVD + mild cognitive dysfunction	17,84±4,30	>0,05
	CVD + severe cognitive dysfunction	16,53±5,30	>0,05
Cleanup workers	CVD + mild cognitive dysfunction	16,74±5,69	<0,1
	CVD + severe cognitive dysfunction	16,60±3,81	<0,02

Thus, cognitive dysfunction associated with immunologic changes in the group of the clean-up workers of the Chernobyl accident could be a result of the direct influence of the accident factors, which creates the conditions for the development of psycho-neurological disorders in the remote period after irradiation. The studies concerning biological mechanisms of postradiation cognitive impairment should be continued [20].

Gene expression studies were performed by RT-PCR. RNA was isolated with QIAcube (QIAGENE, Germany) automatic workstation and NucleoSpin RNAII (Macherey-Nagel, Germany) kit. The High Capacity cDNA Reverse Transcription Kit (Applied Biosystems, USA) was used for cDNA synthesis. RT-PCR was performed on HT-7900 workstation. Gene set included TP53, TP53 I3, BIRC 5, DDB 2, NF-kappa B, TGFR, TERT, TRF1, TRF2, BRCA1, MKNK2, CDKN1B, CDKN2A, CCND1, CDKN1A, FASLG, MADD, BAX, MAPK14, CSF2.

Radiation-exposed with CVD demonstrate over-expression of genes involved in cell entry to the G1 phase of cell cycle and apoptosis – CCND1, TP53, TP53 I3, FasLG. (Figure 4,a). In patients with mild cognitive deficiency an increased TP53 and TP53I3 genes expression was associated with increased CD95+ fraction and decreased numbers of CD4+,

CD4+25+ cells, lower BAX expression and positive correlation between CCND1 gene and bcl-2 expression and radiation dose. Highest expression of CCND1 gene was observed in patients with dementia. In a late period after radiation exposure a relative telomere length shortening, TERF1, TERF2, NFK2B and DDB2 genes hyper-expression (Figure 4b), high oxidative stress markers, low MDA and GSH and an increased T-cell number under G1-cell cycle arrest were demonstrated. BIRC5 and MKNK2 expression was possibly of protective origin, depend on age and stage of cognitive dysfunction.

A positive dependency with radiation dose was demonstrated for CCND1 gene expression (fig 4c), none of such dependencies were shown for other genes. Differences in TERF1 and TERF2 GE were not significant.

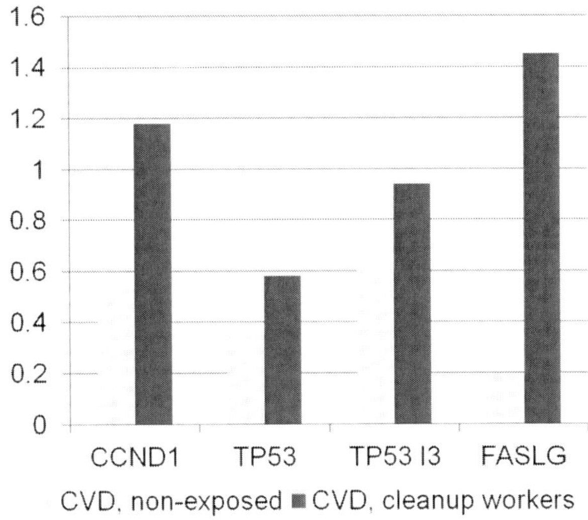

A. Expression of genes, regulating apoptosis.

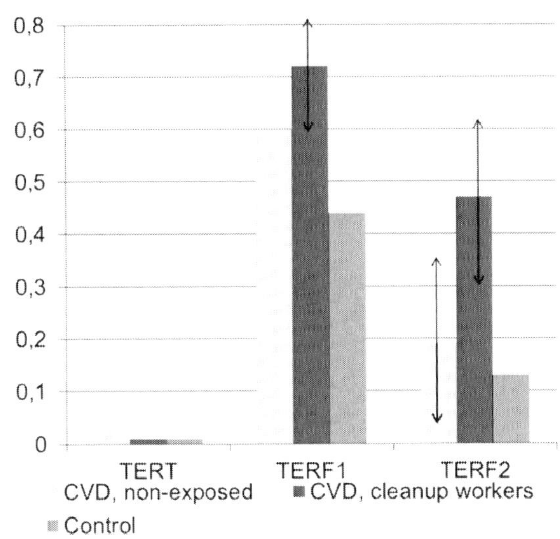

B. Expression of genes, regulating telomere function.

Figure 4. (Continued).

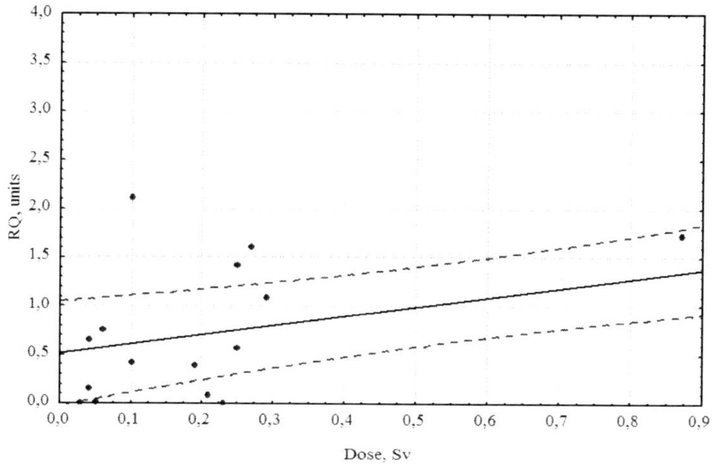

C. Dose dependency for CCND1 gene expression.

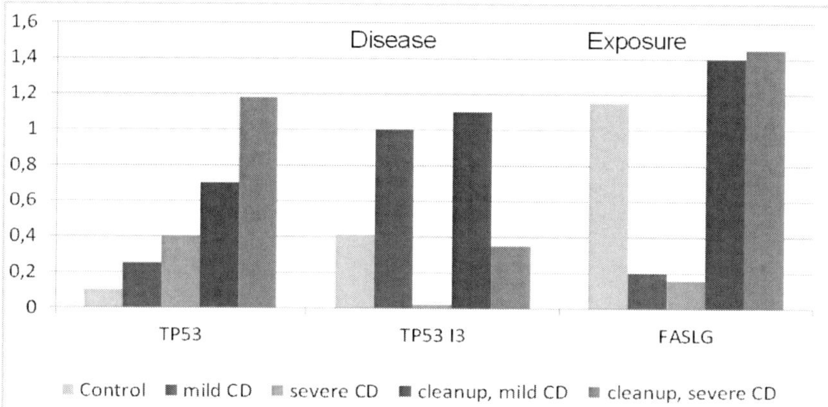

D. Expression of genes due to dose of exposure and disease.

Figure 4. Expression of genes (RQ) in cleanup workers with cerebrovascular disease and cognitive deficit.

Conclusion

RCRM data have demonstrated increased radiation risks of incident cerebrovascular disease, including cerebral atherosclerosis and hypertensive encephalopathy in cleanup workers of 1986-87 exposed over 500 mSv. Ionizing radiation is only one of the set of factors inducing CVD. Genetic predisposition has been identified in 31.7% of the clean-up workers. Body mass excess had been found in 38% of the clean-up workers suffering AH. Smoking habit is revealed in 55.2% of the clean-up workers [2, 3]. Mental health studies have shown elevated rates of cognitive disorders in exposed including depression, suicide ideation and high alcohol abuse rates both in exposed and non-exposed. However, additional evidence is in need.

This study shows a presence of the cellular reactions at a late period after radiation exposure at low and middle dose range. The results allow for expanding knowledge from the experimental data to humans, and testifying the leading role of radiation factor in induction of the effects in immune system in exposed due to Chornobyl accident, and primarily – cleanup workers.

Unlike the experimental data of irradiation the cell lines, the cellular effects in human are modulated not only by a bystander and other non-targeted effects but by a set of confounding factors, from which the most important to our opinion are neuro-immune interaction and lipid metabolism especially the arachidonic acid metabolites [21]. A substantial part of effects is of an adaptive type, like cells transition to the radioresistent phases of cycle with „early" G_1- and „late" G_2-blocks; dose-dependent entry of the immature T- and B-cells to peripheral blood and nonspecific activation. A necessity is to fill the gap and find the link between the molecular changes at low doses and formation of the late non-tumor pathology.

The obtained findings enable the elaboration of strategies of staff selection and follow-up under low-dose exposure. First examples of such approach could be the staff of the 30-kilometer exclusion zone and Chornobyl „Shelter"[22]. On the basis of the experimental data an adaptation to ionizing radiation in doses up to professional limits is supposed for healthy individuals, while at latent tumor processes an increase of genotoxic effects could be possible. It confirms the necessity of the thorougful staff entry checks to the radiation zone. Our data on reverse dependencies between an apoptosis and telomere length are confirmed by the experimental studies [38], in which it was shown that a chemical suppression of telomerase, as a factor which promotes renewal of telomeres length leads to enhanceable apoptosis. So, in humans there is a relationship between violation of apoptosis, telomere shortening, parameters of cellular immunity and radiogenic somatic mutations due to ionizing radiation. At present the dependencies were described between the relative telomere length, dose of irradiation not only over 500 mSv, but at lower doses. Role of age is demonstrated as well as a connection between telomeres, cecorebrovascular disesase and cognitive deficit. Gene expression data show a presence of parameters depending on radiation dose (CCND1, TP53, TERF1, TERF2), presence of cerebrovascular disease and/or accompanying cognitive dysfunction (FASLG), combined action of these parameters (TP53 I3). Other factors like age or presence of increased radiogenic mutations rates (TERT, RTL). Shorter telomeres were demonstrated in groups divided by radiation exposure and cognitive dysfunction. Studies in humans have to be expanded revealing the role of gene regulation after low-dose exposure, as well as a connection between the cognitive function, molecular effects and age at exposure.

REFERENCES

[1] UNSCEAR. Sources and Effects of Ionizing radiation. Volume 2: Effects: Scientific Annexes C, D and E. UNSCEAR 2008 Report. United Nations Scientific Committee on the Effects of Atomic Radiation. United Nations sales publication E.11.IX.3. United Nations, New York, 2011.

[2] Little, M.P., Azizova, T.V., Bazyka, D., et al. Systematic review and meta-analysis of circulatory disease from low-level exposure to ionizing radiation and estimates of the potential population mortality risks *EHP* 120 (11) 1503-1511 (2012).

[3] Buzunov, V., Tereschenko, V., Krasnikova, L., et al. Effects in the Chornobyl clean-up workers p.321-326 in: Health effects of the Chernobyl accident – a quarter of century aftermath (A. Serdiuk et al., eds). *DIA*, Kyiv, 2011.

[4] Buzunov, V., Tereschenko, V., Voychulene, Yu., et al. Non-tumor morbidity and mortality among the Chernobyl clean-up workers 1986-1987. Effect of low-dose ionizing radiation. *Zdrowie i Spoteczenstwo (*1) 101-115 (2011).

[5] Shimizu, Y., Kodama, K., Nishi, N., et al. Radiation exposure and circulatory disease risk: Hiroshima and Nagasaki atomic bomb survivor data, 1950-2003. *BMJ*, 340; b 5349 (2010).

[6] Preston, D., Shimizu, Y., Mabuchi, K. Studies of mortality of atomic bomb survivors. Report 13: Solid cancer and non-cancer disease mortality: 1950-1997. *Radiat. Res.* 160, 381-407 (2003).

[7] Ivanov, V. K., Maksioutov, M.A., Chekin, S., et al. The risk of radiation induced cerebrovascular disease in Chernobyl emergency workers. *Health Physics* 90(3) 199-207 (2003).

[8] Lansdorp, P.M. Telomeres, stem cells, and hematology. Blood 111: 1759-1766 (2008).

[9] Moskalev A.A. Aging and genes. S-Peterburg: *Nauka*, 2008. 191 p.

[10] Regoczy,N, Valent, S, Kormos, L, et al. Telomere length analysis on cord blood cells by the flow-FISH method. *Haematologia* 32: 265-269 (2002).

[11] Gilson, E, Londo c o-Vallejo, A. Telomere length profiles in humans: all ends are not equal. *Cell Cycle* 6: 2486-2494 (2007).

[12] Neuhof, D., Ruess, A., Wenz, F., et al. Induction of Telomerase Activity by Irradiation in Human Lymphoblasts. *Radiat. Res.* 155: 693-697 (2001).

[13] Mathieu, N., Pirzio, L., Freulet-Marriere, M.A. Telomeres and chromosome instability. *Cell Mol. Life Sci.* 61: 641-656 (2004).

[14] Rubio, M.A., Kim, S.H., Campisi, J. Reversible manipulation of telomerase expression and telomere length. Implications for the ionizing radiation response and replicative senescence of human cells. *J. Biol. Chem.* 277: 28609-28617 (2002).

[15] Effros, R.B. Replicative senescence of CD8 T cells potential effects on cancer immune surveillance and immunotherapy. *Cancer Immunol. Immunother.* 53: 925-933 (2004).

[16] Castella, M., Puerto, S., Creus, A., et al. Telomere length modulates human radiation sensitivity in vitro. *J. Toxicol. Lett.* 172: 29-36 (2007).

[17] Willeit¸ P., Willeit, J., Mayr, A., et al. Telomere Length and Risk of Incident Cancer and Cancer Mortality. *JAMA* 304(1) 69-75 (2010).

[18] Scheinberg, P., Cooper, J.N., Sloand, E.M., et al. Association of Telomere Length of Peripheral Blood Leukocytes With Hematopoietic Relapse, Malignant Transformation, and Survival in Severe Aplastic Anemia. *JAMA* 304 (12) 1358-1364 (2010).

[19] Ilyenko, I., Lyaskivska, O., Bazyka, D. Analysis of relative telomere length and apoptosis in humans exposed to ionising radiation. *Exp. Oncology* 33: 235-238 (2011).

[20] Bazyka, D., Loganovsky, K., Ilyenko, I., et al. Cellular immunity and telomere length correlated with cognitive dysfunction in clean-up workers of the Chernobyl accident. *Clinical Neuropsychiatry* 6: 280 – 281 (2013).

[21] Chumak, A., Thevenon, Ch., Gulaya, N., et al. Monohydroxylated fatty acid content in peripheral blood mononuclear cells and immune status of people at long times after the Chernobyl accident. *Rad. Res.* 156 (9) 476-487 (2001).

[22] Health effects of the Chernobyl accident – a quarter of century aftermath (A. Serdiuk et al., eds). DIA, Kyiv, 2011.

In: Biological Responses, Monitoring and Protection … ISBN: 978-1-63463-852-4
Editor: K. P. Mishra

Chapter 7

RADIATION MONITORING AND RADIOPROTECTION IN NUCLEAR DISASTER: LESSONS FROM FUKUSHIMA DAIICHI NUCLEAR POWER PLANT ACCIDENT

Yoshihisa Matsumoto*
Research Laboratory for Nuclear Reactors,
Tokyo Institute of Technology

ABSTRACT

In March 2011, the East coast area of Tohoku district of Japan was attacked by a huge earthquake and succeeding Tsunami, which caused the accident in Fukushima Daiichi Nuclear Power Plant (F1-NPP) leading to a massive release of radioactive nuclides. Nuclear and Industrial Safety Agency (NISA) estimated the released radioactivities of ^{131}I, ^{134}Cs and ^{137}Cs to be 160 PBq, 18 PBq and 15 PBq, respectively. Here I overview, especially focusing on radioiodine, the monitoring system and countermeasures by Japanese government and municipalities and the public exposure and health consequences.

Keywords: Fukushima, radiation, radioiodine, radiocesium, countermeasure

1. THE OUTBREAK OF THE ACCIDENT AND THE RELEASE OF RADIOACTIVITY

On March 11, 2011, at 14:46, an earthquake of magnitude 9.0 occurred in the offshore of Tohoku district. The epicenter was approximately 200 km east of Sendai and the hypocenter was approximately 30 km in depth. This earthquake triggered a tsunami, which reached the height of 40.5 m at the maximum. According to the press release of National Police Agency on June 10, 2012, 15,883 lives were lost and 2,671 people are still not found.

---

* E-mail: yoshim@nr.titech.ac.jp.

There were 6 nuclear reactor units in Fukushima Daiichi Nuclear Power Plant (F1-NPP). Units 4, 5 and 6 were not in operation for planned maintenance. Units 1, 2 and 3 stopped automatically in response to the earthquake and started to be cooled. However, the tsunami of 14 m height submerged the emergency power generators and disabled the cooling of reactors. Then the temperatures and the pressure, as a result of the reactors, gradually rose because of the decay heat. Radioactivity was released into the air by venting operation, to alleviate the pressure, and hydrogen explosion.

Nuclear and Industrial Safety Agency (NISA) estimated the released radioactivities of ^{131}I, ^{134}Cs and ^{137}Cs to be 160 PBq, 18 PBq and 15 PBq, respectively [1]. Nuclear Safety Committee (NSC) estimated the released radioactivities of ^{131}I and ^{137}Cs to be 150 PBq and 12 PBq, respectively [2].

Release of ^{131}I, ^{134}Cs and ^{137}Cs in the Fukushima accident was 9.1%, 39% and 18%, respectively, of the Chernobyl accident (Table 1). The release of strontium and plutonium was still lower in the Fukushima accident than in the Chernobyl accident, probably because the temperature of explosion was lower than boiling point of these elements. It is also noted that most of the radioactivity was blown toward the Pacific Ocean because of the prevailing westerlies.

As a result, area of equivalent level of cesium deposit was more wide-spread in the Chernobyl accident than in the Fukushima accident.

There was an increase in thyroid cancer patients after the Chernobyl accident, probably due to the uptake of radioiodine contained in milk. Therefore, in the Fukushima accident, there is a great public concern on the effects of radioiodine on thyroid. Below I will focus on radioiodine

Table 1. Release of representative radionuclides in Fukushima and Chernobyl accidents

Nuclide	Half Life	Boiling point (^{o}C)	Fukushima (Bq) [1]	Chernobyl (Bq) [3]	Ratio
^{133}Xe	5.25 d	-108	1.1×10^{19}	6.5×10^{18}	1.69
^{131}Io	8.04 d	184	1.6×10^{17}	1.8×10^{18}	0.091
^{132}Te	3.20 d	1390	8.8×10^{16}	1.2×10^{18}	0.076
^{134}Cs	2.06 y	671	1.8×10^{16}	4.7×10^{16}	0.39
^{137}Cs	30.1 y		1.5×10^{16}	8.5×10^{16}	0.18
^{89}Sr	50.5 d	1382	2.0×10^{15}	1.2×10^{17}	0.017
^{90}Sr	29.1 y		1.4×10^{14}	1.0×10^{16}	0.014
^{238}Pu	87.7 y	3230	1.9×10^{10}	1.5×10^{13}	0.0013
^{239}Pu	2.41×10^{4} y		3.2×10^{9}	1.3×10^{13}	0.0002
^{240}Pu	6.54×10^{3} y		3.2×10^{9}	1.8×10^{13}	0.0002

2. Food Control

Ministry of Health, Labor and Welfare (MHLW) enforced "provisional regulation value (PRV)" on March 17, 2011. The food or drinking water, containing radioactivity exceeding this value, were banned for shipping to the market. The value was determined so as to keep the thyroid committed equivalent dose below 50 mSv/y, for iodine, or the committed effective dose below 5 mSv/y, for cesium, assuming that we start eating food with this radioactivity and continue eating for one year.

PRV of radioiodine was initially determined only for drinking water, milk and vegetables, excluding root vegetables and potatoes. As ^{131}I is short-lived, it is considered that food contamination is mainly due to deposition on the surface of vegetables above the ground, rather than uptake from the soil into the plant body. The edible portion of grain is usually inside the hulls. Migration of radioiodine into meat, fish and eggs was considered to be caused indirectly, *i.e.*, through food uptake by animals, and, therefore, took a long time, several times longer than the half life of ^{131}I. However, as a high level of ^{131}I was found in small fish, caught at the offshore, the PRV for fishery product was settled as 2,000 Bq/kg on April 5, 2011.

On April 1, 2012, MHLW enforced a new regulation value so as to keep the committed effective dose below 1 mSv/y. Table 2 summarizes the regulation value before March 31 (PRV) and after April 1, 2012.

Table 2. Regulation value for food and drinking water after Fukushima accident

Category	Before March 31, 2012		After April 1, 2012
	^{131}I	^{134}Cs and ^{137}Cs	^{134}Cs and ^{137}Cs
Drinking water	300 [a]	200	10 [a]
Milk	300 [a]	200	50 [a]
Vegetables	2000[a]	500	100[d]
Grains	-[b]	500	
Meat, Fish, Eggs	2000[c]	500	
Infant foods			50 [a]

Provisional regulation values for uranium and alpha-emitting plutonium and transuranium elements are omitted.

[a] Root vegetables and potatoes are excluded.

[b] Not determined.

[c] Initially not determined. The value was settled for fishery products on April 5, 2011.

[d] Combined into a new category "general foods".

3. Radioiodine Contamination of Foodstuff

Emergency radiation monitoring of foodstuff started on March 16, 2011. Radioactivity measurement was done for 780 samples by March 31, 2011 [4]. Among them, 118 samples were found to contain radioiodine exceeding the PRV: 96 were vegetables and 22 were raw milk. Radioiodine exceeding the PRV was not detected in meat, fish and eggs.

Among vegetables, radioiodine exceeding the PRV was especially frequently detected in spinach. It was also found in other leaf vegetables, with spread leaf. On the other hand, it was found only once in round leaf vegetables, including cabbage. This observation agrees with the assumption that radioiodine contamination would occur mainly due to deposition on the surface.

4. Thyroid Dose

Nuclear Emergency Response Headquarters measured ^{131}I in thyroid glands of 1,149 children aged between 0 and 15 years. Measurement was conducted between March 28 and March 30, 2011 by placing a NaI (Tl) scintillation survey meter on the throat [5]. The objective of this measurement is to rapidly screen the children whose thyroid dose is suspected to exceed 100 mSv. While 0.2 μSv/h is considered equivalent to thyroid gland committed equivalent dose of 100 mSv, >99% showed <0.04 μSv/h and the maximum was 0.07 μSv/h, which is equivalent to 35 mSv. Tokonami et al. measured ^{131}I in thyroid glands using NaI (Tl) scintillation spectrometer between April 12 and April 16, 2011 [6]. The number of residents surveyed was 62 aged between 0 and 83 years. Thyroid gland committed equivalent dose was estimated on the assumption that all radioactivity was uptaken on March 15, 2011 through inhalation or ingestion. According to the "ingestion" scenario, which gives the higher dose estimation than the "inhalation" scenario, the median thyroid gland committed equivalent dose was 3.8 and 4.7 mSv, respectively for over 20 years and 0-19 years populations and the maximum was 37 and 25 mSv.

In the Ukraine cohort, the increase in the odds ratio was statistically significant in the thyroid dose category of 1.00-1.99 Gy and above 2 Gy [7]. In the Belarus cohort, statistically significant increase in the odds ratio was seen in the thyroid dose category of 0.45-0.64 Gy and above [8]. Thus, the thyroid dose in the residents after the Fukushima accident is considered much lower than that in the Chernobyl accident.

5. Thyroid Diagnosis

Thyroid examination started in October 2011. Residents in Fukushima prefecture who were 0-18 years old at March 2013 were subject, based on the consideration that a younger age person might be susceptible to thyroid cancer. In the primary examination, nodules and cysts were screened by ultrasonography. The results were categorized into four classes: A1, no detectable nodules or cysts; A2, nodules with the diameter of ≦5.0 mm or cysts with the diameter of ≦20.0mm; B, nodules with the diameter of >5.1 mm or cysts with the diameter of >20.1 mm; C, based on the overall thyroid status, requiring urgent detailed examination.

As of September 30, 2013, 238,575 people (out of 289,960 subjects, response rate 82.4%) had examination [9] (Table 3). 53.9% were classified into A1 and 45.7% were classified into A2. People diagnosed as class B or class C were 0.7%. There was a public concern that nodules and cysts, albeit small, might have risen from radiation and prodromes of thyroid cancer. Therefore, reference survey was conducted in other areas, *i.e.*, Hirosaki (Aomori), Kofu (Yamanashi) and Nagasaki [10]. The results showed that the frequencies of people diagnosed as class A2 and class B or C were not different from those observed in Fukushima (Table 3).

Table 3. Results of thyroid ultrasonography

	A1	A2	B	C	Total
Fukushima	121,525 (53.9%)	102,453 (45.4%)	1,558 (0.7%)	1 (0.0%)	225,537
Hirosaki	670 (41.1%)	939 (57.6%)	21 (1.3%)	0 (0.0%)	1,630
Kofu	404 (29.6%)	947 (69.3%)	15 (1.1%)	0 (0.0%)	1,366
Nagasaki	779 (56.9%)	582 (42.5%)	8 (0.6%)	0 (0.0%)	1,369

People diagnosed as class B and class C in the primary examination were subject to secondary examination. As of September 30, 2013, 3,459 people have done secondary examination; 58 people were diagnosed or suspected as malignant thyroid tumor by aspiration cytology; 34 people undergo thyroidectomy.

In the Chernobyl accident, increase in thyroid cancer was observed from 1990, *i.e.*, 4 years after the accident [3]. In conjunction with the thyroid dose estimation described above, it is likely that the whole population screening lead to the detection of thyroid cancer without subjective symptoms and that the thyroid cancer cases hitherto observed are not a consequence of radiation exposure in the Fukushima accident, although careful follow-up would be necessary.

6. LESSONS

In the nuclear disaster, great caution should be taken for radioiodine. This is because (1) external and internal exposure at the initial phase is significantly contributed by radioiodine, (2) it accumulates at the thyroid and young children and adolescents are especially susceptible to thyroid cancer and (3) radioiodine rapidly extinguishes by decay and is no longer traceable after extinction. In the case of the Fukushima accident, food control was started promptly. Intensive measurement of milk and spinach appeared effective in the detection of contaminated foodstuff. The whole population screening would result in the increased detection of thyroid cancer cases but these cases might include those not related to radiation. Thyroid measurement promptly after the accident, *i.e.*, before the extinction of ^{131}I, would be critically important for the accurate estimation of thyroid dose.

REFERENCES

[1] Ministry of Economy, Trade and Industry, press release, October 20, 2011, http://www.meti.go.jp/press/2011/10/20111020001/20111020001.pdf (28 March 2014, date last accessed, Japanese).

[2] Nuclear Safety Commission of Japan, press release, April 12, 2011, http://www.nsr.go.jp/archive/nsc/info/20110412.pdf (11 July 2013, date last accessed, Japanese).

[3] United Nations Scientific Committee on the Effects of Atomic Radiation. Annex D. Health effects due to radiation from the Chernobyl accident. *In* Sources and Effects of Ionizing Radiation, UNSCEAR 2008 Report to the General Assembly with Scientific Annexes, Volume II (2008).

[4] Ministry of Health, Labor and Welfare, http://www.mhlw.go.jp/stf/houdou/2r98520000029prx.html (28 March 2014, date last accessed, Japanese).

[5] Akiba S. Epidemiological studies of Fukushima residents exposed to ionising radiation from the Fukushima Daiichi Nuclear Power Plant prefecture—a preliminary review of current plans. *J. Radiol. Prot* 32 (2012) 1-10.

[6] Tokonami S, Hosoda M, Akiba S, et al. Thyroid doses for evacuees from the Fukushima nuclear accident. *Sci. Rep.* 2 (2012) 507.

[7] Zablotska LB, Bogdanova TI, Ron E, et al. A Cohort Study of Thyroid Cancer and Other Thyroid Diseases after the Chornobyl Accident: Dose-Response Analysis of Thyroid Follicular Adenomas Detected during First Screening in Ukraine (1998–2000). *Am. J. Epidemiol.* 167 (2008) 305-312.

[8] Zablotska LB, Ron E, Rozhko AV, et al. Thyroid cancer risk in Belarus among children and adolescents exposed to radioiodine after the Chornobyl accident. *Br. J. Cancer* 104 (2011) 181-187.

[9] Fukushima Prefecture, Prefectural Health Management Division, http://www.pref.fukushima.lg.jp/uploaded/attachment/51093.pdf (28 March 2014, date last accessed, Japanese).

[10] Ministry of Environment, press release, http://www.env.go.jp/press/press.php?serial=16520 (28 March 2014, date last accessed, Japanese).

In: Biological Responses, Monitoring and Protection ... ISBN: 978-1-63463-852-4
Editor: K. P. Mishra

Chapter 8

SOCIAL AND HEALTH EFFECTS OF FUKUSHIMA NUCLEAR ACCIDENT ON RESIDENTS AND EVACUEES

Suminori Akiba*, Athira Nandakumar and Futoshi Uwatoko
Department of Epidemiology and Preventive Medicine
Kagoshima University Graduate School of Medical and Dental Sciences
Sakuragaoka, Kagoshima, Japan

ABSTRACT

The nuclear accident at Fukushima Daiichi Nuclear Power Station (FDNPS) released a large amount of radionuclides into the environment. After the accident, more than 150,000 people were evacuated. Unfortunately, the emergency evacuation was not conducted as timely and efficiently as it should have been. At least 60 elderly who evacuated from hospitals and nursing homes in the area within 20km from FDNPS died by the end of March 2011. During the Fukushima nuclear crisis, a primary concern was the internal exposure of the thyroid gland to radioactive iodine. Stable iodine was not administered by most residents. No extensive effort to measure individual thyroid doses was made, either. By the end of April 2011, when ^{131}I decreased to almost undetectable levels due to its short half-life (=8days), only 1,200 or so residents underwent direct thyroid dosimetry. Another concern was the long-term exposure to radionuclides with relatively long half-lives, such as ^{137}Cs and ^{134}Cs. Fortunately, various surveys have shown that the cumulative radiation doses of residents from those radionuclides will be low. Therefore, the increased risks of malignancies and genetic disorders will be too small to be detected with statistical significance even if the linear non-threshold model is correct, and will probably be buried in the noise of other risk factors, including lifestyles. However, a possible exception is thyroid cancer risk among children as pointed out by the UNSCEAR 2013 Report. It is necessary to examine its risk at least for the next 10 years. The largest health risks after the accident were those related to psycho-social factors rather than the possible direct biological effects of radiation. A lesson from the Fukushima accident is that the people's perception of risk hinges on the trustworthiness of the government, and is not directly linked to the real magnitude of health risk. When

---

* akiba@m.kufm.kagoshima-u.ac.jp.

the government's credibility is questioned, the hyperinflation of perceived health risk takes place.

1. A Brief Summary of the Nuclear Accident

On 11 March 2011, earthquake and tsunami hit the Fukushima Daiichi Nuclear Power Station (FDNPS) of Tokyo Electric Power Company (TEPCO), and crippled its reactor units 1, 2 and 3, which were under operation at that time. At 20:50 JST on March 11, residents within a 2-km radius of the FDNPS were told to evacuate. The evacuation zone was extended to 10 km radius on March 12 morning, and extended further to 20 km radius on that afternoon. The most massive radiation leak from the crippled reactors took place on March 15. At 11:00 on that day, residents in the 20-30 km radius were instructed to take shelter indoors. On that day, heavily radioactive plumes went south in the morning and north-west in the afternoon, and contaminated large areas with radionuclides (Matsumura et al. 2011; Katata G et al. 2011).

There were 1,532 fuel rods in the spent fuel pool which is located 30 meters above the ground in the building of reactor unit 4. This unit had been shut down for shroud replacement and refueling since 29 November 2010. At approximately 6 o'clock on March15 morning, an explosion took place in the 4th floor rooftop area of this unit. According to the Independent Investigation Commission on the Fukushima Nuclear Accident, if the water in the spent fuel pool of reactor unit 4 had been dried up due to this explosion, much larger amounts of radionuclides could have been released into the atmosphere. Fortunately, water remained in the reactor well, which was in the upper part of pressurized vessel, happened to flow into the pool and prevented the dry up (2012).

The current conditions of crippled reactors are not entirely clear. According to TEPCO, the major portion of melted fuel debris in the reactor unit 1 is considered to be in the pedestal or base structure of its container vessel. In the reactor units 2 and 3, the melted fuel debris is in three different places, which are the original reactor core, the plenum of the pressure vessel and the pedestal of its container vessel. TEPCO reported that the temperatures of a primary containment vessel of those reactors are below 50 degrees of Celsius (TEPCO. 31 October 2013).

The nuclear accident in Fukushima is considered to have been caused and exacerbated by several man-made factors. The National Diet of Japan Nuclear Accident Independent Investigation Commission pointed out that the regulatory authorities and the management of TEPCO made self-serving erroneous judgments and missed opportunities to take measures to prevent the nuclear accident (2012). Regarding nuclear safety awareness of TEPCO, Mr. Anegawa, TEPCO executive, admitted that they had problems, and then made the following points: i) awareness was lacking of the need to continuously improve safety, ii) there was a reluctance to improve safety measures beyond regulatory requirements, and iii) the reliability of the existing safety features was overestimated. He also pointed out that the company was reluctant to acknowledge required improvements for fear of losing public confidence in nuclear safety (Anegawa. 2013). On top of that, they were concerned about a possibility that their effort to enhance readiness for severe accidents would be regarded as the de facto acknowledgement of the safety problems in nuclear facilities, which had passed licensing review (Anegawa. 2014). Apparently, our regulatory system could not eliminate their concerns.

2. RADIONUCLIDES RELEASED INTO THE ENVIRONMENT

According to the Nuclear and Industrial Safety Agency (NISA), more than 10 EBq of the radionuclides were released into the atmosphere. Those include ^{133}Xe (11,000 PBq), ^{131}I (160 PBq), ^{132}Te (88 PBq), ^{133}I (42 PBq), ^{134}Cs (18 PBq) and ^{137}Cs (15 PBq). Among radionuclides with long halflives, ^{90}Sr (0.14 PBq), ^{289}Pu (3.2 GBq) and ^{290}Pu (3.2 GBq) were also released into the air (NISA. 20 October 2011).

During the period between June and November 2011, a joint survey of various research institutions, including JAEA (Japan Atomic Energy Agency), collected 11,000 soil samples at 2,200 sites within the 100 km radius of the FDNPS, and determined the radioactivity concentrations of various radionuclides such as ^{137}Cs, ^{134}Cs, ^{131}I, ^{129m}Te and ^{110m}Ag. In addition, major beta emitters, such as ^{89}Sr and ^{90}Sr, and major alpha emitters, including ^{238}Pu and $^{239+240}$Pu, were examined in selected soil samples (N=100). Using the data obtained from those surveys, distribution maps were created (MEXT. 30 August 2011; MEXT. 13 March 2012). The data obtained from the survey are available through the website of JAEA.

It was reported that a large amounts of under-ground water flowed into reactors and spent-fuel pools, and then overflowed and leaked into basements. Into reactor units 1-4, 1000 metric tons of underground water flowed into everyday -- approximately 400 metric tons entered into the building, and the rest (600 metric ton) flowed into the sea after being contaminated by radioactive materials in the trench (METI. 3 September 2013).

Tsumune *et al.* estimated that 3.5 (SD=0.7) PBq of ^{137}Cs were directly released to the sea off the coast of the FDNPS during the period between 26 March and the end of May 2011 (**2012**). In the site of the FDNPS, there are more than 1,000 tanks and other containers storing contaminated water which was used for cooling the damaged reactor units 1-3, and then retrieved in 2013. By early June 2013, approximately 24,000 metric tons of water stored in under-ground tanks was transferred to the above-the-ground tanks in order to avoid unnoticed water leakage. In August 2013, highly radioactive water was found to leak from a storage tank at the facility. The leak is suspected to have started in June. On 19 August 2013, TEPCO submitted the Nuclear Regulation Authority (NRA) a report on the leakage of contaminated water from a water tank in the controlled area at the FDNPS. It is estimated that the amount of leaked water reached approximately 300m^3 (TEPCO. 8 August 2013). After consultations with the International Atomic Energy Agency (IAEA), NRA classified the water leak incident as Level 3 (serious incident) of INES (International Nuclear Events Scale).

On 19 February 2014, TEPCO announced that 110 tons of contaminated water was leaked from the storage tanks (TEPCO. 28 February 2014). The cause of leakage is yet unknown. Repeated water leakage accidents suggest that the safe containment of contaminated water is not an easy task. The corrosion of storage tanks and piping systems is also worried.

3. Evacuation

By the end of 2011, approximately 62,000 people evacuated to the other prefectures (Fukushima Prefecture Government. November 2013). As of July, 2013, approximately 53,000 evacuees from Fukushima Prefecture still stay outside their home prefecture. In addition, 95,000 Fukushima residents are still evacuated from their hometown (Fukushima Prefecture Government. August 2013).

The emergency evacuation was not conducted as timely and efficiently as it should be. At least 60 elderly who evacuated from hospitals and nursing homes in the area within 20km from FDNPS died by the end of March 2011 (the National Diet of Japan Nuclear Accident Independent Investigation Commission. 2012).

Starting 21 April 2011, it was prohibited to enter the 20 km radius zone except for emergency cases. On April 22nd, the Japanese government asked residents to evacuate in approximately 1 month from the areas outside the 20-km no-entry zone in which annual cumulative doses were predicted to exceed 20 mSv. Those areas include Katsurao Village, Namie Town, Iitate Village and a part of Kawamata Town and Minamisoma City. The 20-mSv cutoff line adopted by the government was the toughest of the 20-100 mSv/yr dose range, which was set for the emergency phase in the 2007 recommendation of the International Commission on Radiological Protection (ICRP) (2007). In December 2011, the government decided to lift the evacuation order for areas with residual doses of up to 20 mSv/yr (Cabinet Office 9 October 2013). In the Chernobyl accident, a dose of 5 mSv was used as criteria for relocating residents (Lochard. 2011).

4. Various Measures to Protect Children from Radiation Exposure

On 9 April 2011, Ministry of Education, Culture, Sports, Science and Technology (MEXT) asked municipal school boards in Fukushima Prefecture to restrict the use of school yards with air doses measured at 3.8 microSv/hr or higher (MEXT. 26 July 2012). MEXT estimated that external exposure to 3.8 microSv/hr of radiation for 8 hours outdoors and 1.52 microSv for 16 hours indoors would add up to 20 mSv in a year. – indoor doses were assumed to be 40% of outdoor doses.

In May, the school boards of Koriyama and Minamisoma Cities set limits on the hours spent outdoors while children are at school: 1 hour for class hours and 2 hours after school. This so-called “3-hour” rule was abolished in March, 2012. In August 2011, MEXT announced to abolish its guideline on the use of school yards, and set a new limit to keep the dose received at school less than 1 mSv/yr (MEXT. 26 August 2011).

In response to public concern about the outdoor radiation exposure of children, indoor playgrounds were built at 37 locations in 18 municipalities in Fukushima Prefecture by the end of December, 2012. In addition, many municipalities conducted radiation monitoring of children. In Kawamata Town, for example, simple integrated dosimeters were given to all primary and junior high school students and to all kindergarten children. The average dose during the one-year period between June 2011 and May 2012 was 1.1 mSv after an average natural radiation dose was subtracted. Similar surveys were conducted in other municipalities, including Fukushima City, Nihommatsu Town, Date City, Sohma City and Namie Town. The doses obtained from those surveys were low as was expected.

5. Social Discrimination and Psychological Damage

After the accident, Fukushima people suffered damages caused by harmful rumors. For example, many hotels outside Fukushima Prefecture refused to accommodate people from Fukushima (MHLW. 19 March 2011). In the Kanto Plain area, which includes the Tokyo metropolitan area, gas stations refused to fill gasoline for vehicles from Fukushima Prefecture. In addition, farming and fishing industries in Fukushima Prefecture were seriously damaged by various harmful rumors.

Children were most hurt. Media reported that children evacuated from Fukushima Prefecture were "teased" in various areas. According to the statistics summarized by local legal affairs bureaus across Japan, 711 reports/inquiries regarding human rights violation associated with the earthquake were made by the end of 2012. Human rights abuse reported by the end of 2012 numbered 53. -- It is not possible to identify the cases related to the nuclear accident.

Various episodes of harmful rumors indicate that ordinary Japanese people do not have accurate knowledge about radiation and its health risk. Actually, they have only limited opportunities for learning health risk at school, let alone radiation related health risk. In addition, although the idea of probabilistic effects is a key element for understanding the nature of radiation-induced risk of cancer and genetic disorders, such a concept is foreign to the Japanese, where Buddhism teaches that every result has a cause.

The Japanese government decided to establish education programs for the students of primary schools, middle schools and high schools to learn more about radiation and its health risks at school. In the long run, this kind of efforts to increase the knowledge of the public at large is the most effective measure to improve risk communication in Japan. Such educational programs should eventually provide a large number of teachers with a good understanding of radiation and its associated risks. Note that those teachers can also serve as opinion leaders in various situations, including emergencies (Akiba et al. 2013).

It should also be pointed out that harmful rumors were not limited to inside Japan. In some foreign countries, visitors and products from Japan were refused. International efforts are necessary to eliminate misconception regarding radiation exposure and its health effects.

6. Government Public Relations

In terms of public relations regarding the Fukushima accident, the government seemed to have taken the traditional paternalistic approach, which can be exemplified by the phrase "let the people trust the government; do not give them any information." However, the public was not persuaded in the favor of regulatory and health authorities obsessed with the 100% safety myth. The government's credibility was further damaged by the gradualism exemplified by their belated and indecisive measures after the accident. A lesson from the Fukushima accident is that the people's perception of risk hinges on the trustworthiness of the government and is not directly linked to the real magnitude of health risk. When the government's credibility is questioned, the hyperinflation of perceived health risk takes place.

The public's trust in the government was seriously damaged by the belated disclosure of SPEEDI maps that predicted the patterns of the radiation fallout. The media reported that an office of the Fukushima Prefecture Government received emails containing SPEEDI data from the Japanese government as early as March 12 but the data were not made public until March 23.

The credibility of the government was damaged further in mid-April when MEXT decided to restrict the use of school yards with air doses measured at 3.8 microSv/hr or higher (which corresponds to 20 mSv/yr). Although this guideline was intended to be used only until the end of August 2011, and the cumulative dose could not, therefore, have reached 20 mSv, this decision caused an immediate public outcry because 20 mSv is the maximum dose in the reference level band of 1-20 mSv/yr, as recommended by ICRP for this kind of situation (ICRP. 2007). Protesting the policy of the government, a top advisor to the central government resigned on 29 April 2011. He was reported to have suggested reducing the reference level to 5 mSv/yr.

After the accident, more than 20,000 lecture meetings were held in Fukushima Prefecture. In those meetings, health authorities placed an emphasis on the absence of evidence for an excess health risk associated with exposure to doses below 100 mSv. However, such advice provoked outrage from some participants who pointed out the lack of evidence for the absence of an excess risk in that dose range. This kind of public response can be considered as a backlash against the overemphasis of 100% safety when nuclear projects were promoted.

Public meetings are known to sometimes intensify conflict rather than to resolve controversies (CDC. 2002). This was also true in Fukushima. Lecture meetings were not helpful in promoting public understanding of risk regarding radiation exposure. On the other hand, at small-group meetings, speakers can share information with interested attendees more easily, as was pointed out by a CDC report (2002). Such small meetings are especially useful for informing the public and keeping in touch with community concerns, answering questions, and clearing up any misconceptions or misunderstandings. For people feeling fear and anger, narrative counseling given by empathetic physicians was effective because the public's trust in physicians was still considerable even when the public began to question the trustworthiness of scientists (Chiyo. 2013).

7. External Exposure

In July 2011, the Fukushima Prefecture Government started the Fukushima Health Management Survey. This survey included an effort to estimate individual radiation doses from external exposure. In that dosimetry survey, a self-administered questionnaire was sent to each resident to collect information necessary to evaluate individual doses from external radiation exposure. The number of people to be surveyed is approximately 2 million. As of March 31, 2013, the response rate was approximately 60 % in the areas with relatively high radiation levels but was only 23% in the entire prefecture (Fukushima Prefecture Government. 5 June 2013). Apparently, people in areas with relatively low environmental radiation doses are reluctant to participate in the survey. By the end of 2013, this survey found approximately 155,000 and 1,100 residents to have estimated external doses of 1mSv or higher and of 5mSv or higher, respectively. The highest dose among residents was 25mSv (Fukushima Prefecture Government. 7 February 2014). In many municipalities, individual doses from external exposures among residents were also measured using glass badges. Those surveys also confirmed that individual doses of residents from external exposure were low.

8. Internal Exposure of the Thyroid among Children

In the early days after the accident, a primary concern was the internal exposure of children to ^{131}I, which is known to accumulate in the thyroid and gives the gland relatively high doses among those exposed as young children (Akiba. 2012). After the Chernobyl accident, children in heavy fallout areas had significant internal exposure to the thyroid gland through the consumption of cow's milk which was contaminated with ^{131}I. Fortunately, in the case of Fukushima, radionuclide-contaminated milk was unlikely to have entered the supply chain. From the radiation fallout patterns and other relevant information, the SPEEDI estimated that that the thyroid dose may reach the range of 100-500 mSv for infants spending time outdoors in various areas of Tsushima District of Namie Town during the period between 6:00 on March 12th to 0:00 on April 6th, 2011. Needless to point out, what SPEEDI provides is merely a prediction, and the direct measurements are necessary to know the real doses. Since ^{131}I has the half-life of 8 days, it was necessary to conduct the measurements within a couple of months after the accident. During March 26-30, the Nuclear Emergency Response Headquarters (NERHQ) examined children in Iitate Village, Kawamata Town and Iwaki City, and obtained the thyroid dose data from 1,080 children aged 0-15 years. Only a small portion of children in those three municipalities were examined in this survey. The maximum thyroid dose possibly received by those children was estimated to be 35mSv. It was assumed that the exposure took place from March 12 until the day before the measurement. Although this survey should be considered merely a screening, further and detail examinations were not conducted. Another survey was conducted by Professor Tokonami of Hirosaki University and his colleagues. They estimated that the median thyroid equivalent doses for children and adults were 4.2 mSv and 3.5 mSv, respectively. Using the maximum atmospheric concentration, they estimated that doses for children could exceed 50 mSv (Tokonami et al. 2012). This survey should have been continued and expanded further but was stopped. Regarding the reason why they stopped it, Professor Tokonami told the media that he got a phone call to stop it from a local government officer. A Fukushima Prefecture Government officer denied making such a phone call (Kawakita Shinpo. 21 April 2012; TV Asahi. 11 March 2014). Recently, Professor Tokonami's research team reported different thyroid equivalent dose estimates for residents in Namie Town, using the ^{131}I/^{134}Cs activity ratio which was estimated from their own survey, and the results of whole body counting (WBC) examinations which were carried out by the JEAE a few months after the nuclear accident. The maximum internal exposure of the thyroid to ^{131}I on the basis of the level of ^{134}Cs accumulated in the body measured by the WBC was estimated to be 18 mSv (Hosoda et al. 2013).

The NIRS also estimated thyroid doses on the basis of internal exposure doses of residents and evacuees determined by WBC, in addition to environmental doses obtained from the simulation model of atmospheric dispersion of radionuclides. The median thyroid dose was estimated to be less than 10 mSv even in areas where environmental radiation doses were relatively high. Only 10 % of subjects had doses of 30 mSv or higher (NIRS. 2013).

In May 2012, the WHO published preliminary dose estimation from the Fukushima nuclear accident. The characteristic thyroid doses in the most exposed locations of Fukushima Prefecture were estimated to be within a dose band of 10 to 100 mSv. In one particular location, the assessment indicated that the characteristic thyroid dose to one-year-old infants would be within a dose band between 100 and 200 mSv, with the inhalation pathway being the main contributor to the dose (WHO. 2012). The WHO estimates are evidently higher than those reported by other surveys conducted by Japanese scientists. In 2014, the UNSCEAR reported the estimates of thyroid dose for one-year after the accident. The maximum thyroid dose of 1-year-old infants within the 20-km evacuation zone and the 20-30 km zone were estimated to be 15-82 mSv and 47-83 mSv, respectively. (UNSCEAR. 2014). Difficulties in obtaining reliable thyroid dose estimates are mainly due to the lack of extensive measurement data on radioactive iodine in the thyroid during March and April 2011.

The contribution of radioiodines other than ^{131}I is difficult to evaluate because there is no individual dose measurement. The half-life of ^{132}I is only 2.3 hours and mainly because of the short half-life, the thyroid dose given by this radioactive iodine is considered negligible when compared to ^{131}I. However, in the presence of ^{132}Te in the lung, the dose given by ^{132}I can be comparable to ^{131}I. Fukuda *et al.* determined the activity concentration of radionuclides in the organs of 79 cattle within a 20-km radius around the FNPP. Based on their results, they pointed out the need to pay more attention to ^{132}Te in assessing health risk to the thyroid. According to WHO, 88PBq of ^{132}Te was released into the atmosphere after the accident (WHO. 2012), which is less than one tenth of ^{132}Te released into the environment by the Chernobyl accident (Buzukukov et al. 1993). However, it is difficult to deny the possibility that the plume with highly concentrated ^{132}Te affected relatively densely populated areas. For example, not a small number of evacuees stayed in a small area of Namie Town on March 15.

9. Stable Iodine

The official manual for emergency medicine, which was prepared before the accident, instructs municipality officials to distribute stable iodine together with written instructions to residents and evacuees. The staff members of Fukushima Medical School were provided with stable iodine during the first ten days after the accident (The Asahi Shimbun Degital. 6 November 2013). On 16 March 2011, the NERHQ sent a notice to the municipality offices in 20 km radius zone to give stable iodine tablets at the time of evacuation (Kanatani Y. 2013). In Tomioka and Miharu Towns and in Iwaki City, stable iodine tablets were distributed to 157,000 residents in total without waiting for the instruction of the government (Yomiuri Online. 21 March 2011). In Miharu Town, most of residents aged less than 40 took the stable iodine pill on March 15 (Asahi Shimbun. 2013). On March 21, the NERHQ sent a notice on the use of stable iodine to the mayors of those municipalities and Fukushima Prefecture Governor.

The notice asked them not to leave the decision of administrating stable iodine to residents, and to administer it at the attendance of healthcare professionals. The Nuclear Safety Commission (NSC) admitted that there was a delay in distributing stable iodine (NSC. 2012).

10. Other Internal Exposures

Radionuclides with long half-lives will play a major role in long-term exposure. Those include ^{137}Cs and ^{134}Cs, with half-lives of 30 years and 2 years, respectively. Internal radiation exposure from cesium (^{134}Cs and ^{137}Cs) of residents in Minamisoma City, which is located 23 km north of the FDNPS, was measured using a whole-body counter (Fastscan Model 2250) during the period between 26 September 2011 and 31 March 2012 (Tsubokura et al. 2012). Detection limits were 210 Bq for ^{134}Cs and 250 Bq for ^{137}Cs with a 2-minute scanning. A total of 9,498 residents participated in the study. Cesium was detected in 235 children (16%), and the concentration ranged from 3 to 58 Bq/kg (median, 12 Bq/kg) while 3,051 adults (38%) had detectable levels of cesium, with the concentration ranging from 2 to 197 Bq/kg (median, 11 Bq/kg). In a survey of 1,383 children in Miharu Town, which account for 95% of junior high school pupils in this town, the body burden of ^{137}Cs in the fall of 2012 was below detection limit (300 Bq/body) in all of the participants (Hayano et al. 2013).

In June 2011, the Fukushima Prefecture Government started the examination of internal exposure, using whole body counters (Fukushima Yui net. 2013). Among 118,930 residents and evacuees who underwent measurement by February 2012, 118,904 subjects had doses less than 1 mSv. The subjects with 1mSv, 2 mSv and 3 mSv were 14, 10 and 2, respectively. As of August 2013, the Fukushima Prefecture Government uses approximately 20 WBCs for the internal exposure examination.

11. Food Contamination

Before the accident, the Food Sanitation Act did not stipulate the limit of radionuclide concentrations in foods and drinks produced domestically. On 17 March 2013, Ministry of Health, Labor and Welfare (MHLW) set up provisional regulation limits on them. On 1 April 2012, the MHLW enforced the regulation on food safety to make radiation exposure from ingestion to be 1 mSv/yr or lower (MHLW Food and Drug Administration Bureau).

Since September 2011, MHLW has periodically conducted a food survey using the market basket approach and the duplicate diet method in order to estimate annual radiation doses for various radionuclides through food ingestion. Those surveys showed that the average dose of cesium was much lower than that from radioactive potassium (=0.2mSv/yr), which is naturally originated. In addition, local authorities, including the Fukushima Prefecture Government, have conducted the surveys of food items. Since 2012, all the bags of rice produced in Fukushima Prefecture have been subjected to radiation inspection. Rice with radionuclide concentrations exceeding 100Bq/kg (wet weight) is not allowed to be sold in the market. To date, only 71 bags out of approximately 10 million bags were found to have radionuclides exceeding the limit (Fukushima Prefecture Government. 23 January 2014). In the case of other food items, inspection focuses on food items which are suspected to have relatively high cesium concentrations produced in contaminated areas (MAFF. November 2013).

Private organizations have also made various efforts. Co-op Fukushima (Consumer Co-operative in Fukushima Prefecture) measured radionuclide concentrations in the foods samples obtained from 100 households, using the duplicate diet method during the period between November 2011 and March 2012. Cesium was detected in 10 households, and the maximum value was 6.7 Bq/kg for ^{137}Cs and 5.0 Bq/kg for ^{134}Cs (Sato et al. 2013). In the surveys conducted during the period June-September 2012 and during the period October-December 2012, cesium was detected in 2 and 7 households, respectively. Another important survey has been conducted by Professor Ryugo Hayano of University of Tokyo (The University of Tokyo Foundation. 2013). He has been examining school lunch at preschools in Minamisoma City since January, 2012. To date, radioactive cesium has not been detected (detection limit: approximately 0.5 Bq/kg).

Radio-contamination of fish is also concerned. During the first year after the accident, more than 40% of fish samples were found to have a concentration of 100 Bq/kg-wet or higher (Buesseler. 2012). According to the results of the survey conducted by Fukushima Prefectural Fisheries Experimental Station, more than 90% of demersal fish specimens collected in 2011 had detectable radiocesium (detection limits of ^{134}Cs and ^{137}Cs were 4.70-19.0 and 4.40-19.0 Bq/kg-wet, respectively). In 2012, those figures decreased to three quarters of the 2011 values. The decrease was much more evident in pelagic fish, invertebrates and seaweed. The benthic food web is suspected to be involved in the bioaccumulation of the radioactive cesium detected in demersal fish (Wada et al. 2013). However, since cesium is discharged from fish to seawater through osmotic pressure, bioaccumulation with the magnitudes observed in organic chlorides, such as DDT, is unlikely to happen (Morita. 2011).

Fisheries Agency examines the radionuclide contamination of marine product samples in major fishing ports in Fukushima Prefecture and neighboring prefectures every week (Fisheries Agency 2013). The results (in Japanese) are at the website of Fisheries Agency.

12. Health Effects

Approximately half a year after the accident, the Fukushima Health Management Survey was started. It includes annual health check-ups of residents, the thyroid screening, and a survey of pregnant women and pregnancy outcomes in addition to a survey to estimate the individual radiation dose of residents and evacuees.

The survey of pregnant women and pregnancy outcomes showed that artificial abortion did not appear to increase after the accident (Fukushima Prefecture Government. 12 November 2013). The premature delivery didnot increase, either. The proportion of new-born babies with congenital anomalies was 2.7%, which is similar to the generally known levels (=3-5%).

Before the nuclear accident, the Japan Environment and Children's Study, a nation-wide birth-cohort study involving 100,000 parent-child pairs, was launched to evaluate the impact of various environmental factors on children's health and development. Health outcome surveys and exposure measurements will be continued until the participating children become 13 years old. Before the accident, only a small part of Fukushima Prefecture was included in this study. After the nuclear accident, the entire Fukushima Prefecture was included in the study area.

Recently, Professor Yutaka Hamaoka of Keio University conducted Poisson regression analysis, using the thyroid screening data obtained from the Fukushima Health Management Survey in 30 municipalities in Fukushima Prefecture (Fukushima Prefecture Government. 12 November 2013). He found a statistically significant increase in the frequency of small thyroid nodules (<5 mm) in relation to thyroid dose (Hamaoka. 2013), including the doses estimated by the National Institute of Radiological Sciences (NIRS. 2013) and WHO (2013). The excess prevalence of small thyroid nodule is evident in Namie Town and Iitate Village, and their neighboring areas, which were hit by heavily radioactive plume on March 15 afternoon, 2011. Further studies are necessary to understand the implication of his findings. The radiation doses of Fukushima residents and evacuees are considered to be low. Even so, if linear non-threshold model is correct, cancer risk can be increased among those exposed. The same is also true for genetic disorders. However, even if the model is correct, the increased risk will probably be too small to be detected and will be buried in the noise of other factors, including lifestyles. However, a possible exception is thyroid cancer among residents and evacuees who were exposed to radiation as children (Akiba. 2012). This point was also made by the UNSCEAR report published in 2014 (2014).

To date, there has been no casualty directly caused by radiation exposure. However, the relocation of institutionalized elderly after the accident caused an excess mortality (Yasumura et al. 2013). The changes of lifestyles caused by the accident and subsequent relocation are likely to increase various diseases among residents and evacuees. In addition, the accident seriously damaged social capital of communities, including bonds within the community and between communities. It has been reported that not a small number of residents and evacuees committed suicide after losing their livelihood.

Conclusion

The radiation doses of residents and evacuees from Fukushima Prefecture are considered to be low. Therefore, even if the linear non-threshold model is correct, the increased risks of malignancies and genetic disorders will be too small to be detected and will probably be buried in the noise of other factors, including lifestyles. However, a possible exception is thyroid cancer among children as pointed out by the UNSCEAR report (2014). The health risks caused by social factors turned out to be much larger than the possible direct biological effects of radiation.

A lesson from the Fukushima accident is that the people's perception of risk hinges on the trustworthiness of the government and is not directly linked to the real magnitude of health risk. When the government's credibility is questioned, the hyperinflation of perceived health risk takes place.

REFERENCES

[1] Akiba S. Epidemiological studies of Fukushima residents exposed to ionising radiation from the Fukushima Daiichi Nuclear Power Plant prefecture--a preliminary review of current plans. *J. Radiol Prot.* 2012;32(1):1-10.

[2] Akiba S, Tokonami S, Hosoda M. Summary of discussions at the symposium focusing on problems resulting from the nuclear accident at the Fukushima Daiichi nuclear power plant. *J. Radiol Prot.* 2013;33(1):E1-7.

[3] Anegawa T. Safety enhancement of nuclear power plant and nuclear safety reform in TEPCO. Second nuclear European briefing Budapest, Hungary, 23 September 2013.

[4] Anegawa. Asahi Simbun. 29 March 2014.

[5] Asahi Shimbun. Prometheus's trap. volume 3. *The Asahi Shimbun Company*. 2013. (In Japanese).

[6] Asahi Shimbun Digital. Prometheus's trap. 6 November 2013. (In Japanese) http://digital.asahi.com/articles/TKY201311050626.html?_requesturl=articles/TKY201311050626.html.

[7] Buesseler KO. Fishing for answers off Fukushima. *Science* 2012;338:480-482.

[8] Buzukukov YP, Dobrynin YL. "Release of radionuclides during the Chernobyl accident", The Chernobyl Papers: Dose to the Soviet Population and Early Health Effects Studies (S. Merwin, M. Balonov, Eds), *Research Enterprises*, Richland (1993) 3–21.

[9] Cabinet Office.The Revision of Evacuation Zone. Cabinet Office livelihood support team for nuclear accident evacuees. 9 October 2013. (In Japanese) http://www.meti.go.jp/earthquake/nuclear/pdf/131009/131009_02a.pdf.

[10] CDC. Crisis and Emergency Risk Communication. Center for Disease Control. September, 2002.

[11] Chiyo H. Responses to radiation exposure. *The annual meeting of Japan Public Health Association*. 2013. (In Japanese).

[12] Fukushima Prefecture Government. The progress of the basic survey. The 11th Steering Committee of Fukushima Health Management Survey. 5 June 2013. (In Japanese).

[13] Fukushima Prefecture Government. August 2013. (In Japanese) http://www.reconstruction.go.jp/topics/main-cat1/sub-cat1-4/20130811_kyogikai_shiryo 4-1.pdf.

[14] Fukushima Prefecture Government. November 2013. (in Japanese). http://wwwcms.pref.fukushima.jp/pcp_portal/PortalServlet?NEXT_DISPLAY_ID=U000004&CONTENTS_ID=28571.

[15] Fukushima Prefecture Government. The results of thyroid survey. The 13th Steering Committee of Fukushima Health Management Survey. 12 November 2013. (In Japanese)

[16] http://www.pref.fukushima.jp/imu/kenkoukanri/251112siryou2.pdf.

[17] Fukushima Prefecture Government. The results of survey on pregnant women. *The 13th Steering Committee of Fukushima Health Management Survey.* 12 November 2013. (In Japanese).

[18] Fukushima Prefecture Government. The progress of the basic survey. The 14th Steering Committee of Fukushima Health Management Survey. 7 February 2014. (In Japanese).

[19] Fukushima Prefecture Government. 23 January 2014. http://wwwcms.pref.fukushima.jp /download/1/suiden_syuukei_140123.pdf.

[20] Fukushima Yui Net. The results of internal exposure monitoring using whole body counters obtained by until February 28. 2013. (in Japanese) http://yuinet.beans-fukushima.or.jp.

[21] Hamaoka Y. A Possible Warning from Fukushima: A Preliminary Analysis of Radiation Dose and Occurrence of Thyroid Nodules Using City and Village Level Data" is accepted in the session "Fukushima: Emergency, Dosimetry, Environment, Health. Multidisciplinary European Low Dose Initiative 2013.

[22] Hayano RS, Tsubokura M, Miyazaki M, Satou H, Sato K, Masaki S, Sakuma Y. Internal radiocesium contamination of adults and children in Fukushima 7 to 20 months after the Fukushima NPP accident as measured by extensive whole-body-counter surveys. *Proc. Jpn. Acad Ser. B Phys Biol Sci*. 2013;89(4):157-63.

[23] Hosoda M, Tokonami S, Akiba S, Kurihara O, Sorimachi A, Ishikawa T, Momose T, Nakano T, Mariya Y, Kashiwakura I. Estimation of internal exposure of the thyroid to (131)I on the basis of (134)Cs accumulated in the body among evacuees of the Fukushima Daiichi *Nuclear Power Station accident. Environ Int*. 2013;61:73-6. doi: 10.1016/ j.envint.2013.09.013. Epub 2013 Oct 6.

[24] ICRP. The 2007 Recommendations of the International Commission on Radiological Protection. ICRP Publication 103. Ann ICRP 2007;37:2-4.

[25] Kanatani Y. Public health response to the nuclear accident. *J. Natl. Inst. Public Health* 2013;62:125-131. (In Japanese).

[26] Katata G, Terada H, Nagai H, Chino M. Numerical reconstruction of high dose rate zones due to the Fukushima Dai-ichi Nuclear Power Plant accident. *J. Environ Radioact*. 2012;111:2-12.

[27] Kawakita Shinpo. *Overlooking iodine exposure*. 21 April 2012.

[28] Lochard J. Rehabilitation of Living Conditions After a Nuclear Accident: Lessons From Chernobyl Presentation to the Japanese Cabinet Office Tokyo. 2011 November 28.

[29] Matsumura H. Saito K, Ishioka J, Uwamino Y. Diffusion of radioactive materials from Fukushima Daiichi Nuclear Power Station obtained by Gamma-Ray measurements on expressways. *Japanese Journal of Atomic Energy Society of Japan*. 2011;10(3): 152-162. http://www.mext.go.jp/a_menu/saigaijohou/syousai/1310974.htm.

[30] MAFF. Measures taken at the locale of agriculture, forestry and fisheries. Ministry of Agriculture, Forestry and Fisheries. November 2013. http://www.mhlw.go.jp/topics /bukyoku/iyaku/syoku-anzen/iken/dl/131109-7.pdf.

[31] MEXT. Dose reduction measures in the school buildings and school yards. 26 August 2011. (in Japanese) http://radioactivity.nsr.go.jp/ja/contents/1000/105/27/1000_082614 _1.pdf.

[32] MEXT. Monitoring information of environmental radioactivity level. *Ministry of Education, Culture, Space and Technology*. 30 August 2011. (in Japanese) http://radioactivity. nsr.go.jp/ja/contents/6000/5043/24/11555_0830.pdf.

[33] MEXT. Results of survey on the distributions of radionuclides released from TEPCO Fukushima Daiichi Nuclear Power Plant Ministry of Education, *Culture, Space and Technology*. 13 March 2012. (in Japanese). http://radioactivity.nsr.go.jp/ja/contents /6000/ 5233/24/5233_20120313_20120615_rev20130701.pdf.

[34] MEXT. The record of the measures for the use of school buildings and school yards. 26 July 2012. http://www.mext.go.jp/component/a_menu/other/detail/__icsFiles/afieldfile/2012/07/26/1323887_10.pdf.

[35] METI. The overview of measures against the water contamination in FDNPS. Ministry of Economy, *Trade and Industry*. 3 September 2013. (In Japanese) http://www.meti.go.jp/earthquake/nuclear/pdf/osensuitaisaku_houshin_ref01.pdf.

[36] MHLW. Instruction to hotel industries regarding the accommodation of Fukushima nuclear accident evacuees. Section of Life Higien, Ministry of Health, Labor and Welfare. 19 March 2011. (In Japanese). http://www.mhlw.go.jp/stf/houdou/2r98520000014tr1-img/2r98520000015j0k.pdf.

[37] MHLW. On the estimated intake of radionuclides from food consumption. Ministry of Health, Labor and Welfare. 17 April 2013. (In Japanese) http://www.mhlw.go.jp/shinsai_jouhou/dl/20130417-3.pdf.

[38] MHLW Food and Drug Administration Bureau. Measures against radionuclides in food and the current status. Department of Food Safety, Food and Drug Administration Bureau, Ministry of Health Labor and Welfare. 25 October 2013. http://www.mhlw.go.jp/shinsai_jouhou/shokuhin.html.

[39] Morita T. Seafood and radionuclides. Organization for the Promotion of Responsible Tuna Fisheries. News letter June 2011.

[40] NSC. Draft summary of proposal for preventive administration of stable iodine. 30th meeting of subcommittee for health care and management for radiation exposure and medical care, Nuclear Facility Expert Committee, Nuclear Safety Commission 7 February 2012. (In Japanese) http://www.nsr.go.jp/archive/nsc/senmon/shidai/hibakubun/hibakubun030/siryo4-1.pdf.

[41] NIRS. The results of dose estimation for internal exposure in early phase after the accident of Fukushima Daiichi Nuclear Power Plant. Fukushima Nuclear Disaster Monitoring Survey of Ministry of Environment in 2012. National Institute of Radiological Sciences. February 2013. (in Japanese)

[42] NISA. Errata for the data on the amounts of released radionuclides. News Release. Ministry of Economy, Trade and Industry. 20 October 2011. (in Japanese)

[43] http://www.meti.go.jp/press/2011/10/20111020001/20111020001.pdf NIRS. The results of dose estimation for internal exposure in early phase after the accident of Fukushima Daiichi Nuclear Power Plant. Fukushima Nuclear Disaster Monitoring Survey of Ministry of Environment in 2012. National Institute of Radiological Sciences. February 2013. (in Japanese).

[44] Sato O, Nonaka S, Tada J. Intake of radioactive materials as assessed by the duplicate diet method in Fukushima. *J. Radiol. Prot.* 33 823 doi:10.1088/0952-4746/33/4/823

[45] TEPCO. Measures against contaminated water in the sea. Tokyo Electric Power Company. 8 August 2013. (in Japanese). http://www.tepco.co.jp/nu/fukushima-np/roadmap/images/c130808_07-j.pdf

[46] TEPCO. Cooling conditions of reactor unit 1-3 containment vessels of FDNPS. Tokyo Electric Power Company. 31 October 2013. (in Japanese) http://www.meti.go.jp/earthquake/nuclear/pdf/131031/131031_01f.pdf

[47] TEPCO. The submission of a "nuclear power plant accident report" on the leakage from TEPCO contaminated water storage tanks, RO concentrated water receiving tank (H6

area C1 tank), to the NRC. 28 February 2014. (in Japanese) http://www.tepco.co.jp/cc/press/2014/1234564_5851.html

[48] The Independent Investigation Commission on the Fukushima Nuclear Accident. Research Investigation Report. Rebuild Japan Initiative Foundation. March 2012. (in Japanese)

[49] The National Diet of Japan Fukushima Nuclear Accident Independent Investigation Commission. 2012.

[50] The University of Tokyo Foundation. Activity report from Professor Ryugo Hayano. 2013. (in Japanese). http://utf.u-tokyo.ac.jp/result/project-result/5.html 2013

[51] Tokonami S, Hosoda M, Akiba S, Sorimachi A, Kashiwakura I, Balonov M. Thyroid doses for evacuees from the Fukushima nuclear accident. *Sci Rep*. 2012;2:507. doi: 10.1038/srep00507.

[52] Tsubokura M, Gilmour S, Takahashi K, Oikawa T, Kanazawa Y.Internal radiation exposure after the Fukushima nuclear power plant disaster. *JAMA*. 2012;308(7):669-70.

[53] Tsumune D, Tsubono T, Aoyama M, Hirose K. Distribution of oceanic 137Cs from the Fukushima Dai-ichi Nuclear Power Plant simulated numerically by a regional ocean model. *J. Environ Radioact*. 2012;111:100-8.

[54] TV Asahi. Hodo (News) station. 11 March 2014.

[55] UNSCEAR 2013 Report. Rport to the General Assembly. Vol I. Scientific annex A. Levels and effects of radiation exposure due to the nuclear accident after the 2011 great east-Japan earthquake and tsunami. United Nations. 2 April 2014.

[56] Wada T, Nemoto Y, Shimamura S, Fujita T, Mizuno T, Sohtome T, Kamiyama K, Morita T, Igarashi S. Effects of the nuclear disaster on marine products in Fukushima. *J. Environ Radioact*. 2013;124:246-54.

[57] WHO. Preliminary dose estimation from the nuclear accident after the 2011 Great East Japan earthquake and tsunami. World Health Organization. 2012.

[58] WHO. Health risk assessment from the nuclear accident after the 2011 Great East Japan Earthquake and Tsunami based on a preliminary dose estimation. *World Health Organization*. 2013.

[59] Yasumura S, Goto A, Yamazaki S, Reich MR. Excess mortality among relocated institutionalized elderly after the Fukushima nuclear disaster. Public Health. 2013;127(2):186-8.

[60] Yomiuri Online. 21 March 2011. (in Japanese). http://www.yomiuri.co.jp/national/news/20110321-OYT1T00020.htm

In: Biological Responses, Monitoring and Protection … ISBN: 978-1-63463-852-4
Editor: K. P. Mishra

Chapter 9

THE CHALLENGING ISSUE OF HIGH LEVELS OF BACKGROUND RADIATION AND CANCER INCIDENCE IN RAMSAR, IRAN

S. M. J. Mortazavi and K. P. Mishra
Shiraz University of Medical Sciences, Iran

SUMMARY

Some inhabited areas in Ramsar, Iran are known for extraordinary levels of natural background radiation. People in these high background radiation areas (HBRAs) are continuously exposed to high levels of indoor and outdoor gamma radiation as well as high concentrations of indoor Rn-222. In spite of current controversies, some of the studies performed so far have shown increase in the frequency of chromosome aberrations in the lymphocytes of residents of HBRAs. Howevere, no repeatable detrimental effects caused by high levels of natural radiation in Ramsar have been detected so far. Furthermore, the issue of carcinogenesis of high levels of natural radiation in these areas is not proven yet. As the indoor Rn-222 level in Ramsar reaches 31 kBq m^{-3}, Mortazavi et al. previously collected the overall cancer mortality, lung cancer mortality and neonatal death rate in different districts of Ramsar. This study surprisingly showed that the highest recorded rate of lung cancer mortality was in an area with normal levels of indoor radon, while the lowest lung cancer mortality rate was in another area where the highest concentrations of radon in the dwellings were found. Recently, an expert team supervised by Mortazavi studied the tumor markers in the inhabitants of HBRAs of Ramsar. In this study, the serum levels of PSA, CA15.3, CA125, Cyfra21-1, CEA, CA19.9, AFP and Tag72 tumor markers in 35 residents of HBRAs and 35 individuals from a normal background radiation area (NBRA) were measured. Furthermore, the level of background gamma radiation and indoor radon concentration in the houses of all participants were measured. The findings of this study did not show any significant difference between the mean levels of PSA, CA15.3, CA125, CA19.9 and AFP of the residents of HBRAs and NBRAs. However, the levels Cyfra21, CEA and Tag72 in HBRA group were significantly higher compared to those of NBRAs. On the other hand, a statistically significant correlation between the external gamma radiation dose or indoor radon level and the concentration of CEA, Cyfra-21 and TAG 72 biomarkers was observed. As the mean external exposure (indoor and outdoor) in HBRAs of Ramsar is about 6 mSv y^{-1} for a population of about 2000 residents, to obtain statistically significant data in epidemiological studies in the mSv range, large sample sizes are needed. Therefore, to overcome the limitations of sample size in epidemiological studies, we believe that studies with relevant tumor markers can shed some light on the dark corners of the challenging issue of cancer incidence in high background radiation areas.

In early days of life the background radiation level was much higher than what exists today [1]. The level of natural radiation more than 3.5 billion years ago, when the living organisms appeared on the Earth, was about 3 times higher than its current level. Also in the early days of life, it is believed that there were as many as 100 million natural reactors, such as found in Oklo, Gabon. The dose rate around these natural reactors was estimated to be up to 47 Gy per minute [2]. It has been reported that the natural background radiation levels on Earth vary by at least two orders of magnitude. Therefore, all living organisms are continuously exposed to a wide range of background radiation levels. Areas with high levels of background radiation are found in Yangjiang, China; Kerala, India; Guarapari, Brazil; and Ramsar, Iran [3]. According to UNSCEAR report Ramsar has some inhabited areas with the highest known natural radiation levels in the world [4]. The mean and maximum annual natural terrestrial radiation doses to the inhabitants of Yangjiang, Kerala, Guarapari and Ramsar are indicated in Table 1.

Table 1. Mean and maximum annual natural terrestrial radiation doses to the inhabitants of some areas around the world

Country	Area	Approximate population	Absorbed Dose rate in air[a] (nGy h^{-1})
Brazil	Guarapari	73 000	90-170 (street) 90-90 000 (beaches)
Iran	Ramsar[b] Mahallat	2 000	70-17 000 800-4000
India	Kerala	100 000	200-4 000
China	Yangjiang	80 000	370 (average)

[a] Includes cosmic and terrestrial radiation.

[b] It should be noted that the monazite sand beaches at Guarapari in Brazil have a higher dose rate, but these areas are uninhabited. Therefore it can be claimed that Ramsar has the highest level of natural radioactivity studied so far.

Source: UNSCEAR 2000 [4].

The extraordinary high levels of natural background radiation in Ramsar seem primarily to be due to the high concentrations of radium-226 and its decay products originated from hot springs and also high levels of thorium in rocks and soil (travertine deposits) [5]. As soil is derived from the process of weathering of local bedrocks, the radioactivity levels in local soils and the foods grown in these soils are also high. Figure 1 shows a personal gamma radiation dosimeter (DKG-21 ECOTEST Card) that is placed on very hot soil samples collected from high background radiation areas of Ramsar.

Figure 1. A personal gamma radiation dosimeter (DKG-21 ECOTEST Card) is placed on hot soil samples collected from high background radiation areas of Ramsar.

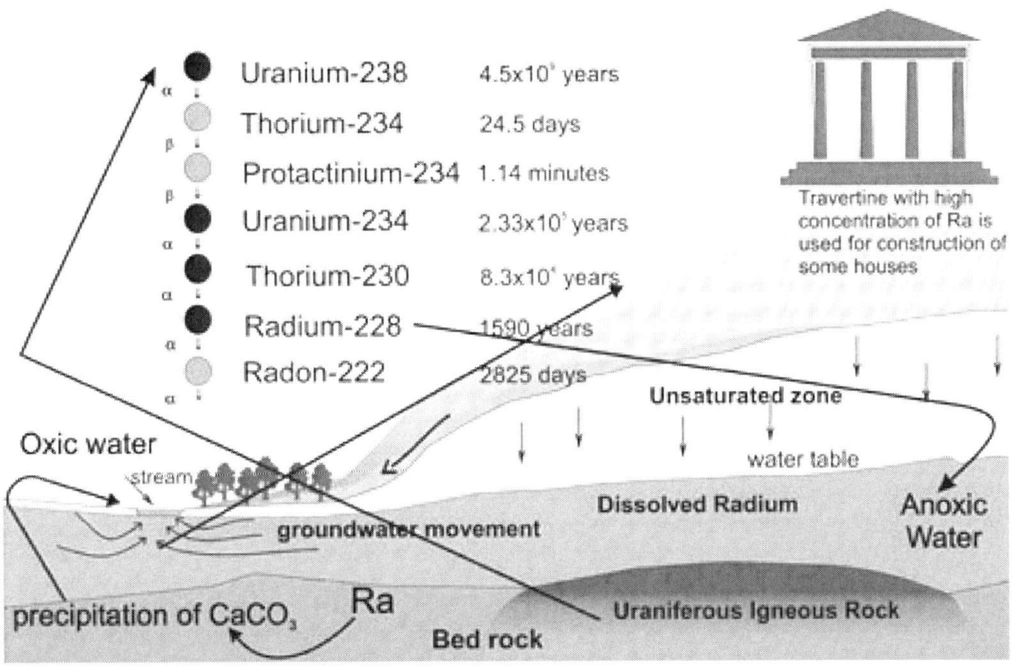

Figure 2. The origin of high levels of natural radioactivity in Ramsar. (Originally from Mortazavi et al. 2001 reproduced in Introduction to Medical Geology by C. B. Dissanayake, Rohana Chandrajith).

The origin of high levels of natural radioactivity in high background radiation areas (HBRAs) of Ramsar is shown in Figure 2. As it is indicated in the figure, high concentrations of uranium in igneous bedrocks cause high levels of radium-226. Although uranium is not soluble in anoxic ground water, when it decays into radium-226, it becomes soluble in ground water. Groundwater carries the dissolved radium to the surface, passing through pores and fractures in the rock. When underground water reaches the surface at hot spring locations, calcium carbonate precipitates out of solution and radium-226 substitutes for calcium ($RaCO_3$). Therefore, high concentrations of radium carbonate can be usually found in the residue of hot springs. Interestingly, in some cases the residents of the hot areas used the radium-enriched rock from the hot springs as building materials to construct their houses.

Figure 3 shows bedroom walls of a house in HBRAs of Ramsar. The owner had used the sedimentary stones formed at hot springs for construction of his house. As shown in the figure, two survey meters show the dose rates of 142 and 143 μGy/h on the wall of the bedroom of this dwelling. Considering the recommendations of the Iranian National Radiation Protection Department, the owners moved to a new house and nobody lives in this house now.

Figure 3. Bedroom walls of a house in HBRAs of Ramsar. The owner had used the sedimentary stones formed at hot springs for construction of his house.

As a natural laboratory for radiation biology, Ramsar has always been the center of attention of radiobiologists, health physicists and environmentalists (Figure 4). It has been reported that the annual radiation absorbed dose in HBRAs of Ramsar is extraordinary high, reaching to 260 mSv that is 13 times higher than the annual dose limit of 20 mSv suggested for radiation workers by ICRP. Due to the very high public exposures in HBRAs of Ramsar, some experts have recently suggested that an effective remedial action program is urgently needed [6]. Howevere, no repeatable detrimental effects caused by high levels of natural radiation in Ramsar have been detected so far [3].

a

b

Figure 4. Ramsar has always been the center of attention of radiobiologists, health physicists and environmentalists. (a) A scientist is measuring the dose rate around the hot orange trees. (b) A group of scientists from Korea, Japan, India, USA and Iran are visiting HBRAs of Ramsar.

The risk of exposure to low doses of ionizing radiation has been globally politicized due to a variety of reasons. This has led to a frequently exaggerated perception of the health effects of low dose radiation and to lasting public controversies [7, 8]. In spite of current controversies, some of the studies performed on the health effects of exposure to elevated levels of natural radiation so far have shown increase in the frequency of chromosome aberrations in the lymphocytes of residents of HBRAs [9-11]. However, it has been reported that the role of elevated levels of natural radiation in HBRAs is less significant than smoking in induction of stable-type chromosome aberrations (translocations) [12].

The issue of carcinogenesis of high levels of natural radiation in these areas is not proven yet. As the indoor Rn-222 level in Ramsar reaches 31 kBq m^{-3}, Mortazavi et al. previously collected the overall cancer mortality, lung cancer mortality and neonatal death rate in different districts of Ramsar. This study surprisingly showed that in contrast with what was expected from the linear no-threshold (LNT) model, the highest recorded rate of lung cancer mortality was in an area with normal levels of indoor radon, while the lowest lung cancer mortality rate was in another area where the highest concentrations of radon in the dwellings were found (Figure 5).

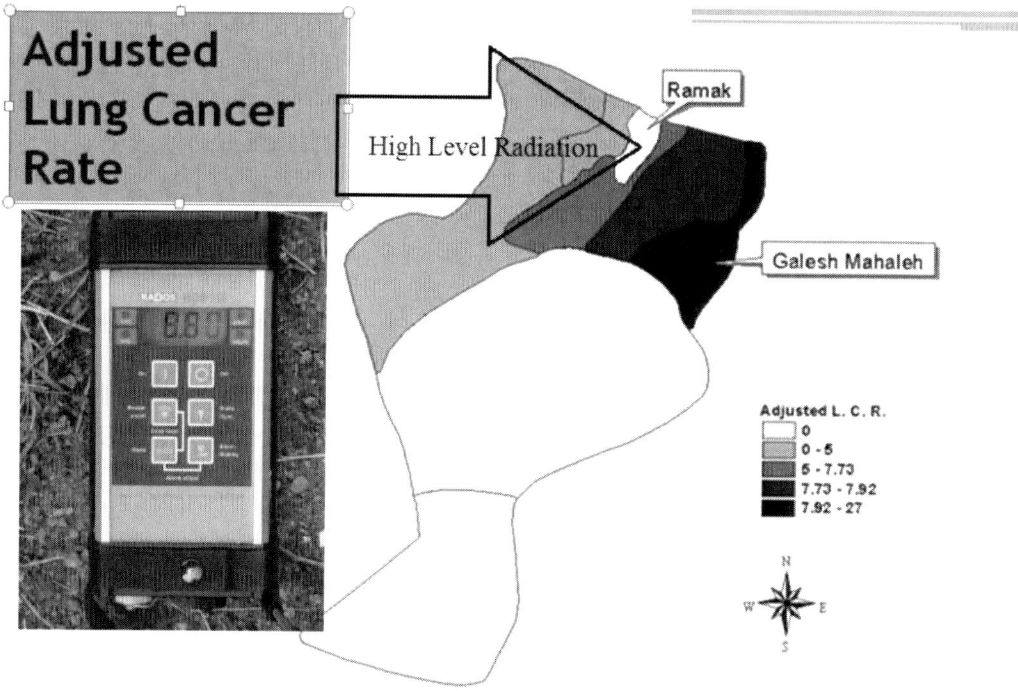

Figure 5. Adjusted lung cancer mortality rate in different districts of Ramsar. The highest recorded rate of lung cancer mortality was in an area with normal levels of indoor radon, while the lowest lung cancer mortality rate was in another area where the highest concentrations of radon in the dwellings were found.

Studies performed in other HBRAs did not show a significant relationship between cumulative radiation dose and cancer rate. In a study conducted by Tao et al. in 2012, the mean cumulative radiation doses from natural radiation in the high background radiation and control area residents were 84.8 and 21.6 mGy, respectively. This study indicated that the mortality rate due to leukemia or solid cancers was not related to the residents' cumulative radiation doses [13]. On the other hand, Nair et al. studied the relationship between the background radiation and cancer incidence in Kerala. According to the findings of their study, no cancer site was significantly related to cumulative radiation dose in site-specific analysis. Furthermore, leukemia was not significantly related to high background radiation, too [14].

Recently, an expert team supervised by Mortazavi studied the tumor markers in the inhabitants of HBRAs of Ramsar. In this study, the serum levels of PSA, CA15.3, CA125, Cyfra21-1, CEA, CA19.9, AFP and Tag72 tumor markers in 35 residents of high background radiation area (HBRA) and 35 individuals from a normal background radiation area (NBRA) were measured [15]. Furthermore, the level of background gamma radiation and indoor radon concentration in the houses of all participants were measured. The findings of this study did not show any significant difference between the mean levels of PSA, CA15.3, CA125, CA19.9 and AFP of the residents of HBRAs and NBRAs. However, the levels Cyfra21, CEA and Tag72 in HBRA group were significantly higher compared to those of NBRA. From the other point of view, a statistically significant correlation between the external gamma radiation dose or indoor radon level and the concentration of CEA, Cyfra-21 and TAG 72 biomarkers were observed.

On the other hand, Heidari et al. also measured the levels of three different tumor markers; namely CEA, PSA and CA19-9; in the blood serum of inhabitants of the high background radiation areas of Ramsar [16]. They compared their results to those of a control group from Noshahr (Nowshahr), a city 85 km far from Ramsar with a non-homogeneous population [17]. Heidari et al. reported a significant increase in the levels of PSA and CA19-9 markers of Ramsar group compared to that of the control group ($p < 0.001$). They could not observe any difference in CEA levels. In spite of these positive results, they reported that the Ramsar residents were in good health (they were examined by a physician). Moreover, the high number of positive markers in Noshahr was considerable. Therefore, future study is needed to further validate this result and to determine the level of positivity to tumor markers in both cities. Some of the shortcomings of this paper (three fatal methodological problems and the mistake in interpretation of the phenomenon of adaptive response) are discussed by Mortazavi and Mozdarani [18].

As the mean external exposure (indoor and outdoor) in HBRAs of Ramsar is about 6-10 mSv y^{-1} for a population of about 2000 residents, to obtain statistically significant data in epidemiological studies in the mSv range, large sample sizes are needed. Therefore, to overcome the limitations of sample size in epidemiological studies, we believe that studies with relevant tumor markers can shed some light on the dark corners of the challenging issue of cancer incidence in high background radiation areas.

REFERENCES

[1] Karam PA, Leslie SA: Calculations of background beta-gamma radiation dose through geologic time. *Health physics* 1999, 77(6):662-667.

[2] Jaworowski Z: Beneficial effects of radiation and regulatory policy. *Australas Phys Eng Sci Med* 1997, 20(3):125-138.

[3] Mortazavi SMJ, Mozdarani H: Non-linear phenomena in biological findings of the residents of high background radiation areas of Ramsar. *International Journal of Radiation Research* 2013, 11(1):3-9.

[4] UNSCEAR: Sources and Effects of Ionizing Radiation. In., vol. 2000 Report to the General Assembly with Scientific Annexes. New York:United Nations: United Nations Scientific Committee on the Effects of Atomic Radiation (UNSCEAR); 2000.

[5] Ghiassi-Nejad M, Mortazavi SMJ, Cameron JR, Niroomand-Rad A, Karam PA: Very high background radiation areas of Ramsar, Iran: Preliminary biological studies. *Health physics* 2002, 82(1):87-93.

[6] Sohrabi M: World high background natural radiation areas: Need to protect public from radiation exposure. *Radiation Measurements* 2013, 50(0):166-171.

[7] Kellerer AM: Risk estimates for radiation-induced cancer--the epidemiological evidence. *Radiation and environmental biophysics* 2000, 39(1):17-24.

[8] Karam PA, Mortazavi SMJ: The Very High Background Radiation Area in Ramsar, Iran: Public Health Risk or Signal for a Regulatory Paradigm Shift? *Kjell Andersson* 2001:495-502.

[9] Hayata I, Wang C, Zhang W, Chen D, Minamihisamatsu M, Morishima H, Wei L, Sugahara T: Effect of high-level natural radiation on chromosomes of residents in southern China. *Cytogenetic and genome research* 2004, 104(1-4):237-239.

[10] Zakeri F, Rajabpour MR, Haeri SA, Kanda R, Hayata I, Nakamura S, Sugahara T, Ahmadpour MJ: Chromosome aberrations in peripheral blood lymphocytes of individuals living in high background radiation areas of Ramsar, Iran. *Radiation and environmental biophysics* 2011, 50(4):571-578.

[11] Ghiassi-Nejad M, Zakeri F, Assaei RG, Kariminia A: Long-term immune and cytogenetic effects of high level natural radiation on Ramsar inhabitants in Iran. *J Environ Radioact* 2004, 74(1-3):107-116.

[12] Zhang W, Wang C, Chen D, Minamihisamatsu M, Morishima H, Yuan Y, Wei L, Sugahara T, Hayata I: Effect of smoking on chromosomes compared with that of radiation in the residents of a high-background radiation area in China. *Journal of radiation research* 2004, 45(3):441-446.

[13] Tao Z, Akiba S, Zha Y, Sun Q, Zou J, Li J, Liu Y, Yuan Y, Tokonami S, Morishoma H *et al*: Cancer and non-cancer mortality among Inhabitants in the high background radiation area of Yangjiang, China (1979-1998). *Health physics* 2012, 102(2):173-181.

[14] Nair RR, Rajan B, Akiba S, Jayalekshmi P, Nair MK, Gangadharan P, Koga T, Morishima H, Nakamura S, Sugahara T: Background radiation and cancer incidence in Kerala, India-Karanagappally cohort study. *Health physics* 2009, 96(1):55-66.

[15] Taeb S, Mortazavi SMJ, Ghaderi A, Mozdarani H, Kardan A, Almeida CE, Soleimani A, Mortazavi SAR, Haghani M, Soofi A: Alterations of PSA, CA15.3, CA125, Cyfra21-1, CEA, CA19.9, AFP and Tag72 tumor markers in human blood serum due to long term exposure to high levels of natural background radiation in Ramsar, Iran. *International Journal of Radiation Research* 2014, 12(2):133-138.

[16] Heidari MH, Porghasem M, Mirzaei N, Mohseni JH, Heidari M, Azargashb E, Movafagh A, Heidari R, Molouki A, Larijani L: The effect of high level natural ionizing radiation on expression of PSA, CA19-9 and CEA tumor markers in blood serum of inhabitants of Ramsar, Iran. *Journal of Environmental Radioactivity* 2014, 128(0):64-67.

[17] Mehra H: A demographic study of some populations in shahrestan nowshahr, mazandaran, Iran, with reference to the genetic structure. Doctoral thesis, 1984: Durham University; 1984.

[18] Mortazavi SMJ, Mozdarani H: PSA, CA19-9 and CEA tumor markers in blood serum of inhabitants of Ramsar, Iran, Heidary et al., Journal of Environmental Radioactivity. *Journal of Environmental Radioactivity* 2014, 132: 121-122.

In: Biological Responses, Monitoring and Protection … ISBN: 978-1-63463-852-4
Editor: K. P. Mishra

Chapter 10

THE EFFECTS OF ELECTRON BEAM AND GAMMA IRRADIATION CONDUCTED IN AIR ON THE PROPERTY BEHAVIOUR OF SEMICRYSTALLINE MEDICAL GRADE POLYURETHANE

Kieran A. Murray[1], James E. Kennedy[1], Brian McEvoy[2], Olivier Vrain[2], Damien Ryan[2], Richard Cowman[2], and Clement L. Higginbotham[1,*]

[1]Materials Research Institute, Athlone Institute of Technology, Dublin Road, Athlone, Co. Westmeath, Ireland

[2]Synergy Health, IDA Business & Technology Park, Sragh, Tullamore, Co. Offaly, Ireland

ABSTRACT

Gamma and electron beam radiation would be considered as an attractive method of sterilisation for medical grade polyurethane, due to the radiation resistance provided by the material. However, extremely high irradiation doses can alter the structure and/or function of macromolecules, resulting in oxidation, chain scission and crosslinking. This chapter reviews various topics related to the radiosterilisation of polymers with regards to the technologies used, the effects of radiation on polymers and the mechanisms related to modification and degradation processes that have been reported in the literature. A concise history outlining the development of polyurethane materials is explored; while the behaviour of this material under radiation conditions will be discussed. The importance of environmental conditions and in particular the temperature of the polyurethane material during radiation exposure is highlighted. A detailed account of the behaviour of the material in terms of the thermal, structural, mechanical and surface characteristics will be analysed using experimental data. While it is clear following critical appraisal that irradiated virgin polyurethane will not suffice as an acceptable material for medical device applications, there is methods that yield a high probability for

[*]Tel: 00353 90 6424463; fax: 00353 90 6424493; e-mail address: chigginbotham@ait.ie

the successful stabilisation of polyurethane material to reduce/eliminate radiation degradation.

Introduction

Sterilisation is essential for all devices/products whose end use is anticipated to infringe on the body's defence mechanisms, to come into contact with damaged tissue or to be implanted into the body. It has become increasingly complex due to the requirements of preventing patients from the exposure of infections caused by viable organisms on the instruments/devices during their care. Unsatisfactory sterilisation of medical devices may result in significant institutional costs as a consequence of patient nosocomial infections and mortality/morbidity concerns [1]. Many medical devices and components manufactured from polymers tend to deform when exposed to thermal sterilisation temperatures, yet the devices and components have good capabilities resisting the temperatures reached by irradiation [2]. Ethylene oxide (EtO) gas is the most popular method of sterilisation in the medical device industry, however, the industry is undergoing drastic changes on account of the strict regulations trend for ethylene oxide gas, which require zero emissions and high speed treatment. Accordingly, recent attentions are turned to irradiation which provides the superiority to meet the aforementioned desires [3].

The main objective of sterilisation via radiation is to minimise over-dosing and/or under-dosing the product, while maximising microorganism death under acceptable conditions in order to prevent any undesirable changes such as degradation effects in materials. Depending on the sterilising technique used, materials can either act physically or chemically leading to alterations of the structure or function of macromolecules which can result in chain scission, oxidation, crosslinking, melting and hydrolysis [4, 5]. With regards to radiosterilisation of medical grade polymers such as polyurethane, this can lead to undesirable changes in the polymeric material where the prediction and control of the properties are sometimes not well understood. Polyurethane elastomers, generally, are resistant to high energy radiation, however, this is dependent on the number of aromatic rings and the concentration of ester groups in the polymer [6].

1. Key Radiosterilisation Techniques

An extensive range of sterilisation technologies exists in the medical device industry and these include electron beam radiation, gamma radiation, X-Ray radiation, UV, ethylene oxide, steam, dry heat and low temperature hydrogen peroxide gas plasma. Each of these technologies has a diverse array of features which can have a desirable or undesirable effect depending upon the polymer and the application for end use. The ideal outcome of sterilisation is to reduce the bioburden to a minimum while ensuring no modifications occur to the physical and optical properties of the product. This section will focus on three primary methods of radiosterilisation for the medical device industry which are as follows:

- X-Ray
- Gamma ray
- Electron beam

1.1. X-Ray

X-Ray sterilisation of medical devices was first introduced more than 40 years ago where theoretical and experimental studies were conducted by the accelerator manufacturers, US National Bureau of Standards and universities. Although the commercial use of X-Ray sterilisation for medical device applications has been adopted for approximately 15 years, uptake of the technology has been relatively slow as a result of low output power of early accelerators. Due to the recent advancements of high-power, high-energy accelerators, it is now possible to achieve desirable dose uniformity ratios and this is quickly changing the landscape of radiation technology [7]. X-Ray dose rates are between the high dose rates of electron beam irradiators and the low dose rates of gamma ray sources which are suitable for applications that are sensitive to dose rate. The choice of energy source for a particular application is usually based on the practical aspects of the irradiation process, such as material thickness and density, throughput rate, dose uniformity and economics.

Essential characteristics of an X-Ray irradiator improve as the energy increases and these include an increase in the power conversion efficiency, a decrease in the angular dispersion and an increase of the penetration in processed materials [8]. In terms of the technical advantages for higher energies, it improves the dose uniformity, reduces the exposure time and improves the power utilisation. Furthermore, it increases material output, reduces treatment costs and betters the product quality in a practical sense. X-Ray sterilisation is now a strong competitor to traditional technologies such as gamma ray, EtO and electron beam processing [9]. This provides research and development in the medical device industry with an alternative method of sterilisation for their current and next generation products.

X-Rays are produced by introducing a metal target between the electron beam and the product to be treated. To improve the electron to photon conversion, these X-Ray target materials are made of high atomic number (high-*Z* target material) metals. Water cooled tantalum or tungsten is preferred for large area targets [10]. The diagram presented in Figure 1 illustrates the electron to proton converter.

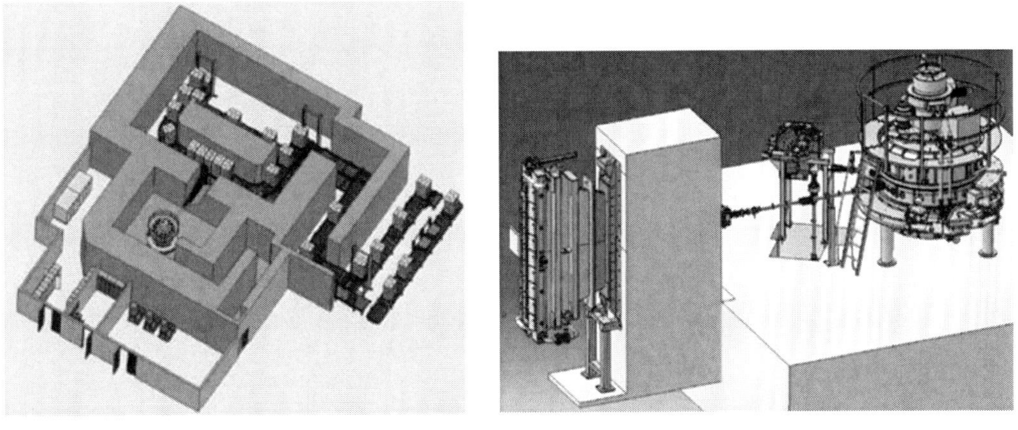

Figure 1. Image to the left: typical layout of an industrial size X-Ray facility. Image to the right: Rhodotron accelerator. (Images provided by Synergy Health.)

1.2. Gamma Ray

The principle industrial application of gamma radiation lies with the destruction of microorganisms. The earliest work that was carried out on this technology was in the 1940's where the effectiveness of irradiation as an industrial scale microbial process was evaluated [12]. Effects of gamma radiation on microorganisms were well established and subsequently it was thought that two gamma sources would have practical industrial potential for this purpose. The two sources under discussion are caesium 137 (^{137}Cs) and cobalt 60 (^{60}Co) which were considered due to the relatively high energy of their rays and the fairly long half-life (30.1 years for ^{137}Cs and 5.27 years for ^{60}Co) [13]. Caesium 137 is formed in spent fuel rods and was thought that it would be the better option to use. However, this choice of source did not last due to the high expense of the separation process required to make sufficient amounts available for industrial gamma processing. Seeing that ^{60}Co can be produced by the exposure of naturally occurring ^{59}Co to neutrons, it became the predominant source. The absorption of one neutron per atom converts ^{59}Co into radioactive ^{60}Co [2]. These gamma ray emitters became a popular radiation source for medical and industrial applications.

In recent times, the use of electron accelerators as a provider of a radiation source (and sometimes equipped with an X-Ray converter) is gradually increasing. Nonetheless, gamma irradiators are almost impossible to replace, particularly for non-uniform and high density products. To date ^{60}Co is used almost solely as a gamma radiation source for industrial use, mainly because of its easy production process and its non-solubility in water [13]. The primary application for gamma radiation has become the sterilisation of medical devices, with ^{60}Co remaining the predominant source. Expansion of this technology was quite slow at the initial stages, where only 55 plants were in operation across the world in 1975 with an average source strength of 36000 Ci [14]. The Food and Drug Administration (FDA) initially classified irradiation as a lengthy validation procedure for medical devices, as the approval for batch release into the market place was very slow. This situation eased away in the 1970's with growing distress over the probable carcinogenicity of ethylene oxide (an alternative cold sterilisation process available for heat sensitive plastics) and the number of irradiators doubled to 110 between 1975 and 1984 with the average source also doubling to 625000 Ci. By 1985, 133 industrial gamma irradiators were in operation throughout the world, with 90 % of them dedicated to medical devices [14, 15]. For over fifty years, gamma radiation has been recognised as an extremely safe, cost-competitive methodology for the sterilisation of healthcare products, components and packaging. With the technology being compatible with single-use disposable medical devices, gamma radiation is being used by an ever increasing percentage of the healthcare industry. Over 200 gamma radiation facilities are in operation worldwide with 120 of the plants located in Europe and the USA [16, 17].

Sterilising and decontamination of products by gamma radiation can be a sophisticated process requiring a great deal of knowledge in such areas as the selection of materials, process control and the kinetics of microbial inactivation. The energy emitted for the photons of gamma radiation is transferred to electrons in the target material yielding highly active electrons and highly reactive free electrons. With regards to sterilisation, it is this energy that is used to destroy the reproductive cells of microorganisms by destroying DNA chains which in turn prevents replication. Figure 2 demonstrates a typical layout of a gamma radiation facility. Gamma radiation can be compared to other methods of sterilisation with regards to the advantages and disadvantages of sterilising medical devices. There are two particular advantages that exist for the gamma radiation process and these are penetration and reliability. Gamma radiation has very good reliability as the process solely depends on one parameter in contrast to ethylene oxide which has at least four parameters i.e. gas concentration, temperature, humidity and time [2]. In addition, the process offers low temperatures and moisture levels for sensitive materials that cannot tolerate such conditions. However, gamma irradiation can have undesirable effects on various material characteristics post processing, as a result of free radical reactions [2].

At the current time, gamma irradiators used for sterilisation are still under big demand. Nearly half of all medical devices are sterilised today by using gamma radiation which have a ^{60}Co radioactive source. There is no sign of this market declining at the moment. However, with ^{60}Co price fluctuations, limited availability, increasing regulations on transportation and difficulty related to the disposal of radioactive sources, this may lead medical device manufacturers to evaluate alternative sterilisation techniques [18].

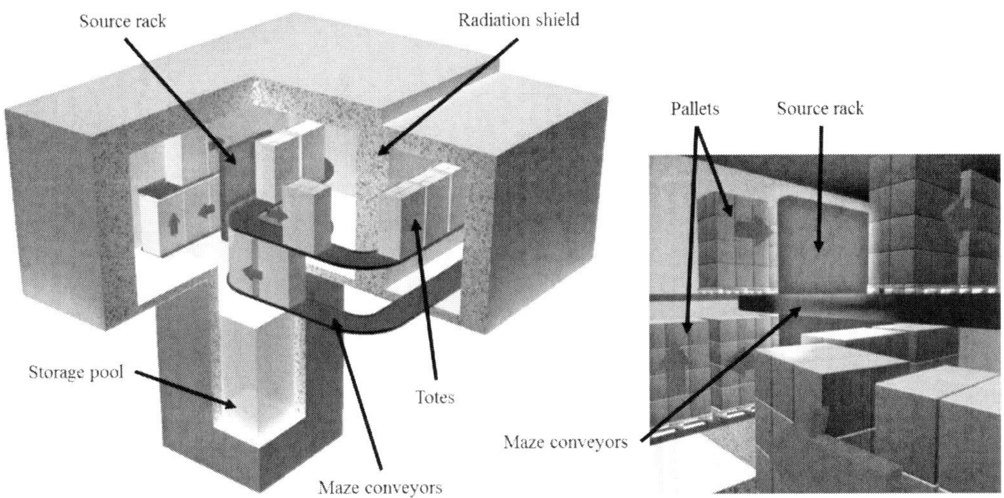

Figure 2. Schematics of typical gamma cells tote plant (left) pallet facility (right). (Images provided by Synergy Health.)

1.3. Electron Beam Irradiation

Electron Beam processing first started in the 1950's, but has only grown significantly in the last twenty five years. This is due to the enhancement of accelerator technology which brought lower cost per kilowatt, advances in radiation chemistry and the availability of polymer formulations designed for the process [19, 20]. Electron beam radiation is an attractive process, mainly because the throughput rates are very high and the treatment costs per unit of product are often competitive with more conventional chemical processes [21]. At present, radiation processed products are being used in health care, food preservation and polymer based industries. The process has provided added value to numerous products of more than 85 billion dollars [22-24].

Electron beam radiation is a form of ionising energy which is normally classified by its low penetration and high dosage rates. The beam (pulsed or continuous) consists of a highly charged, concentrated stream of electrons which are generated by the acceleration and conversion of electricity. Electrons are produced by equipment referred to as accelerators [20]. All electron beam accelerators have some features in common and work on the same principle as a television tube: 1.) Electrons are generated by a heated filament which forms the electron gun. 2.) A voltage gradient draws the electrons away from the gun and accelerates them through the vacuum tube. 3.) As the high voltage beam of electrons passes from the beam tube and through the scan magnet, an oscillating magnetic field sweeps the beam back and forth across the scan window. A diagram showing the operating principles of the accelerator can be seen in Figure 3, while a typical layout of an electron beam facility is illustrated in Figure 4.

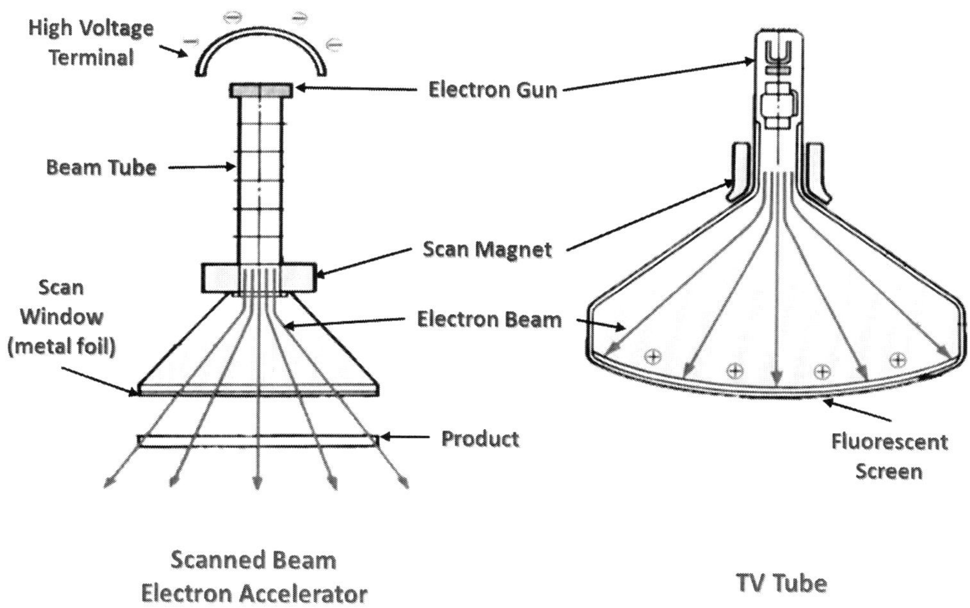

Figure 3. Electron beam operation principles [25].

Electron beam accelerators are used in diverse industries to enhance the physical and chemical properties of materials and to reduce undesirable contaminants, such as pathogens and/ or toxic by-products. This system is attracting more attention recently for the sterilisation of medical devices due to the many advantages as controllability, flexibility in operation, no emissions, high speed processing, favourable economics and the ability to switch off the source [26]. Sterilisation with this technology is achieved by passing the product in front of the electron beam which absorbs the electrons. This absorption of energy alters various chemical and biological bonds within the product. It is this absorption of energy that destroys the reproductive cells of microorganisms by destroying their DNA chains [20].

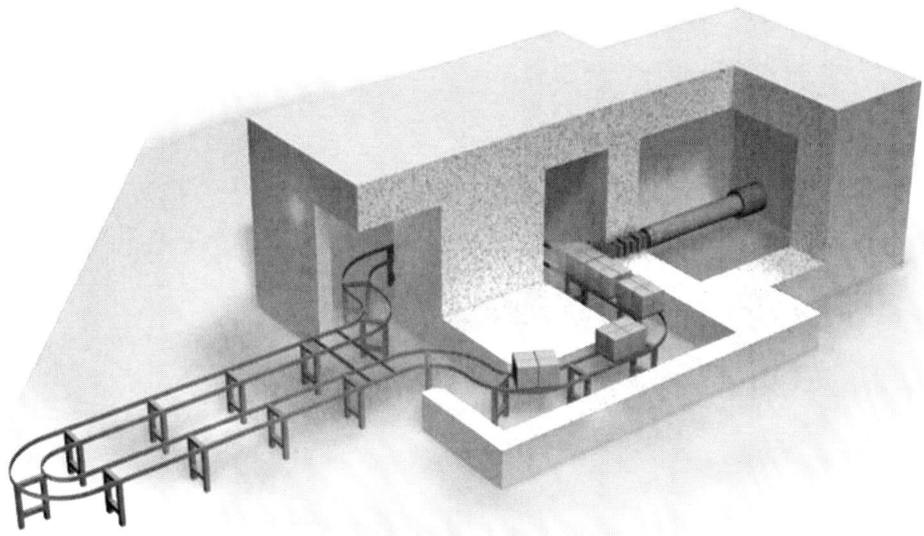

Figure 4. Typical layout of an electron beam irradiator. (Image provided by Synergy Health.)

1.3.1. Energy and Power

Beam energy defines the penetration of a beam and hence the radiation field within the beam axis while the beam current defines the number of electrons per second which penetrate the product or irradiation volume and hence the dose rate [27]. In the Systeme International (SI), gray is a unit of measure for the energy of ionising radiation that is absorbed by a mass. One gray is equal to 1 joule per kilogram and 1 kilogray (kGy) is equal to 1 joule per gram, (J/g). The rate at which radiation energy is delivered defines the power of the radiation source. For an electron beam source: Power (Watts) = Average beam current (amps) x Average beam energy (Volts). The power determines the rate at which product can be processed and the voltage determines the depth of penetration [28].

1.3.2. Penetration and Dose

Dose rate is the term used for the amount of ionising radiation delivered per unit time during an irradiation process. The dose rate (D_R) for an electron accelerator can be written in terms of beam current (I) and irradiation field area (R):

$$D_R = KI/R \tag{1}$$

Where K is the stopping power of electrons, which is dependent on the energy of electrons and density of materials being irradiated [29]. Interaction of high energy electrons with condensed materials is dependent on both the kinetic energy of the electrons and atomic composition of the irradiated materials. The intensity of the absorbed dose first increases and then decreases as can be seen from Figure 5. Absorbed dose is basically the amount of energy deposited into a specified mass of material, where the unit is in gray (Gy) and is more frequently called kilogray (kGy). The absorbed dose for electron beam processing can be determined by the following: 1) assign dose from Monte Carlo probability codes and area through-put equations or 2) infer dose using a surrogate (a dosimeter or dose meter) that has proven reliability and is stable with respect to irradiation response or one that relates to the materials that are actually being processed [19].

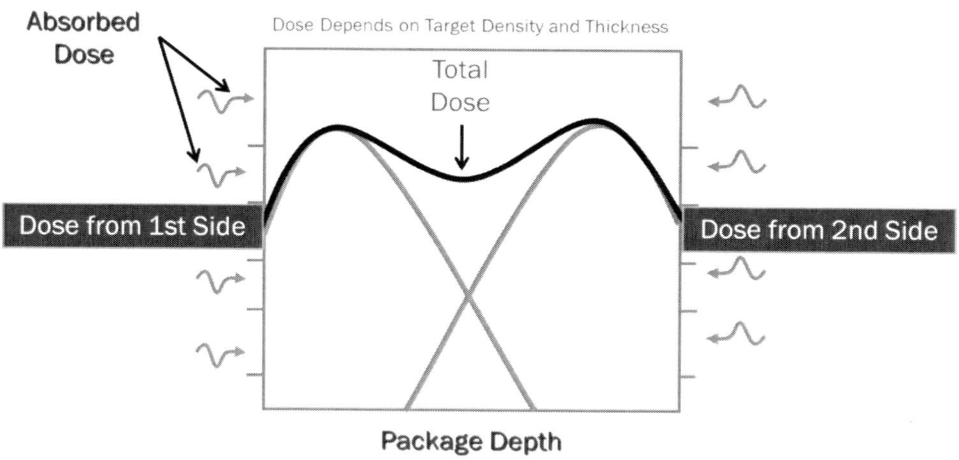

Figure 5. Variation of absorbed dose with depth and the effect of irradiation from both sides.

Table 1. The advantages and disadvantages of radiosterilisation techniques

Method	Advantages	Disadvantages
X-Ray	❖ Full pallet processing ❖ Can power off source ❖ High penetration ❖ Excellent dose uniformity ratio ❖ Very labour efficient ❖ Powered by electricity ❖ Faster treatment than gamma irradiation	❖ Longer processing time than electron beam ❖ Inefficient energy conversion ❖ Material degradation - radiation stable materials required
Gamma ray	❖ It is generally a colder method than electron beam ❖ It has a high sterility assurance level ❖ Control of the method is very easy - can be achieved by the parameter of applied dose ❖ High density materials can be processed ❖ Flexible batch size ❖ No quarantine period ❖ Good penetration ❖ Non-uniform packing	❖ Long processing time ❖ Material degradation - radiation stable materials required ❖ Dose rate is lower than electron beams and X-Ray ❖ It has no dose rate flexibility
Electron beam	❖ Provide short exposure times ❖ Faster cycle time - time is money ❖ Flexible batch size ❖ Simplifies validations ❖ Can power off source ❖ Eliminates operator dependencies ❖ Enable real time monitoring ❖ Control of the method is very easy - can be achieved by the parameter of applied dose ❖ More material compatibility than gamma ray	❖ Limited penetration ❖ Uniform packing required ❖ Material degradation - radiation stable materials required

1.4. Evaluation of Radiation Sterilisation Techniques

Radiosterilisation processes can be used for the treatment of pharmaceuticals and medical devices, however; it is hard to assign a particular method as each of them has their own advantages and disadvantages. Therefore, the sterilisation process should be chosen based on the physical and chemical structure of the product being treated. The superiority of radiation to EtO and other sterilisation methods are known all over the world which helps to understand the relatively fast increase of the constitution of irradiation institutions. Radiation technology is very efficient in terms of time and cost for the treatment of materials in the polymer industry, sterilisation of packaging material for medical devices, pharmaceuticals and food industry. In particular from a pharmaceutical perspective, sterilisation in the terminal step (final packed drug) is the most important advantage of radiation sterilisation [30]. A comprehensive evaluation of each of the radiosterilisation techniques is illustrated in Table 1, which details the advantages and disadvantages respectively.

2. Effects of Radiosterilisation on Polymers

The greatest disadvantage of radiosterilisation is the deleterious effects that irradiation has on numerous polymers that would otherwise be the preferred choice of material for manufacture. During the irradiation process, radicals are formed in the material in less than a nanosecond. Subsequently, complex mixtures of ionised and excited species are created in the matrix of the polymer causing the species to react and transfer energy through several mechanisms such as dissociation, addition and abstraction. All processes, particularly the reactions with oxygen can last for a number of seconds, hours, days, or even months in the polymer and can result in a number of physical and chemical modifications [31]. The process of interaction for high energy electrons with material can be divided into three primary events.

- *Ionisation*: Here, the fast electron transfers its energy to the bonding electron in the absorbing material and the electron is knocked out. This event only takes place when the transferred energy during the interaction is higher than the bonding energy of the bonding electron.

  (i) $AB \rightarrow AB^+ + e^-$

  The ionised molecule separates into a free radical ion at approximately the same time.

  (ii) $AB^+ \rightarrow A\bullet + \bullet B^+$

- *Excitation:* This causes the movement of electrons from the ground state to the excited state.

  (iii) $AB \rightarrow AB^*$

  Eventually the excited molecule dissociates into free radicals:

  (iv) $AB^* \rightarrow A\bullet + B\bullet$

- *Capture of Electron:* This is also an ionisation process where the molecules can capture electrons with lower energy. As a result the ion can dissociate into a free radical and a radical ion:
  (v) $AB + e^{-} \rightarrow AB^{-}$
  (vi) $AB^{-} \rightarrow A\bullet + \bullet B^{-}$

Apart from the primary reactions, a number of secondary reactions occur in which ions or excited molecules take part. The end result from the events illustrated above is that through the diverse primary and secondary fragmentations, radicals are produced that can initiate a free radical process leading to polymerisation, cross-linking, backbone or side-chain scissions, structural rearrangements and so on [32]. These molecular changes within the polymer result in the following changes: embrittlement, loss of elongation, recrystallisation, increased free radical content and peroxide levels, acid formation or pH shifts, consumption of radiation stabilising additives, generation of volatiles and extractable changes in hardness, colour, density, molecular weight, flexural properties, modulus and tensile strength etc.

The three fundamental responses of polymeric materials to electron beam are (Figure 6 demonstrates these three responses):

1) Crosslinking – the formation of an insoluble material
2) Neutral – little to no effects on mechanical properties
3) Scissioning – the lowering of the molecular weight of a material

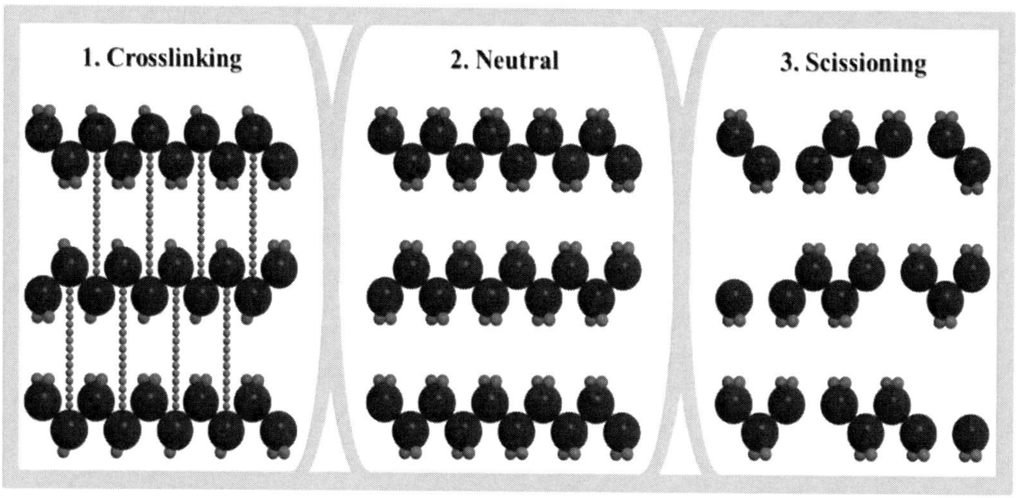

Figure 6. Three principal responses of post irradiation effects on polymeric material.

2.1. Crosslinking and Chain Scission

Polymers usually undergo simultaneous scission and crosslinking but in most cases one or the other clearly predominates. The structure of polymers often decides the preponderance of one process over another [29]. Crosslinking is where intermolecular bonds link one polymer chain to another. It is basically the formation of three dimensional networks which occur simultaneously. The degree of crosslinking is proportional to the radiation dose [33]. Crosslinking can enhance the mechanical, thermal and chemical properties, environmental and radiation stabilities. Chain scissioning is usually an undesirable effect, especially for medical polymers, as it can reduce polymer molecular weight and degrade mechanical properties. In addition, the chain scission process reduces both strength and stiffness, lowers fatigue and the ability to elongate and weakens adhesive bonds. This process is initiated when polymer radicals react with oxygen to form peroxy radicals that subsequently chain propagate, chain branch, and eventually terminate the chain reaction [31]. The sensitivities of various polymers towards the crosslinking and chain scission process can be evaluated from their "G" values which describe the yield of radiation induced reactions. A "G" value is the number of molecules/ions produced or destroyed per 100 ev energy absorbed. In terms of crosslinking and chain scission, the "G" value is written as G(X) and G(S) respectively [29]. Polymers that are affected by either process are presented in Table 2.

Table 2. Classification of polymers with regards to their radiation response [32]

Polymers predominantly crosslinking	Polymers predominantly degrading
Polyethylene	Polyisobutylene
Polypropylene	Poly(α-methylstyrene)
Polystyrene	Poly(vinylidene chloride)
Poly(vinyl chloride)	Poly(vinylidene fluoride)
Poly(vinylidene fluoride)	Polychlorotrifluoroethylene
Poly(vinyl alcohol)	Polyacrylonitrile
Poly(vinyl acetate)	Polyvinylbutyral
Copolymer of ethylene and vinyl acetate	Poly(methyl methacrylate)
Poly(vinyl methyl ether)	Polymethacrylonitrile
Polybutadiene	Poly(methacrylamide)
Polychloroprene	Polyoxymethylene
Copolymer of styrene and acrylonitrile	Poly(propylene sulfide)
Copolymer of styrene and butadiene	Poly(ethylene sulfide)
Natural rubber	Cellulose
Chlorinated polyethylene	Polyalanine
Chlorosulfonated polyethylene	Polylysine
Polyamides	DNA
Polyesters (PLLA, PGA, PCL)	
Polyurethanes	
Polysulfones	
Polyacrylates	
Polyacrylamides	
Polydimethylsiloxane	
Copolymer of vinylidene fluoride and	
hexafluoropropylene	
Copolymer of ethylene and tetrafluoroethylene	
Polydimethylphenylsiloxane	
Phenol-formaldehyde resins	
Urea-formaldehyde resins	
Melamine-formaldehyde resins	

2.2. Discolouration

Discolouration is another potential negative result from the irradiation of polymers especially for the medical device markets. This is developed from specific chromophores during the stabilising radiochemistry reactions which generally causes the material to turn a yellowish colour [34]. A colour change is mainly attributable to the presence of phenolic additives, which are contained in the formulation to inhibit oxidation by alkyl and peroxy radicals produced during irradiation or exposure to high temperatures. The phenolic antioxidants are normally phenols and during oxidation they generate stable phenoxyl radicals. These phenoxyl radicals can either dimerise, disproportionate or continue to react with other polymer radicals [3].

2.3. Temperature

Another negative issue that can occur is the rise in temperature of the material during irradiation. An absorbed dose of 10 kGy is equal to 2.4 calories per gram, which can increase the temperature of water by 2.4 °C [35]. Most polymeric materials have heat capacities lower than water and exhibit correspondingly a higher temperature rise [36]. In general, the rate of chemical reactions increase with temperature as a result of the greater proportion of molecules which have energies in excess of the activation energy, and this may apply to the radiation induced reactions in polymers [36]. Below the glass transition temperature of polymers, a large number of stable radicals are generated and trapped in the glassy state. A rise of 10 °C doubles the reaction rates [25]. Table 3 presents the temperature rise in common materials.

Table 3. Temperature rise in common materials [25]

Material	ΔT in °C/kGy
Water	0.24 °C
Polyethylene	0.43 °C
Polypropylene	0.52 °C
Polyvinylchloride	0.75 °C
Aluminium	1.11 °C
Copper	2.63 °C

2.4. Molecular Changes

Ionising radiation leads to molecular changes in polymers and these are commonly characterised using a G factor which is used to quantify the yield of an event [37]. Such events may include molecular alterations, gel content and changes to the molecular weight. The molecular weight of a polymer and its distribution can be significantly modified due to the exposure of ionising radiation. High energy treatments such as electron beam irradiation result in free radical generation, main chain scission and crosslinking, and subsequently alter the distribution of chain size in the bulk polymer [38].

2.5. Oxidation

When ionising radiation is conducted in an air atmosphere, it can lead to severe structural degradation of the polymer. Most polymers including those that are prone to crosslinking in an inert environment, will likely undergo chain scission when oxygen is available during ionisation [39]. Time dependency of the radiation process plays a major role with free radical reactions which can initiate severe radiation degradation of polymers, particularly in the presence of oxygen because of high diffusion, mobility and reactivity. Free radical reactions in radiation induced polymer oxidation are as follows [38]:

$R + \gamma \rightarrow R^{\bullet}\}$ initiation
$R^{\bullet} + O_2 \rightarrow RO_2^{\bullet}\}$ propagation
$RO_2^{\bullet} + RH \rightarrow RO_2\,H + R^{\bullet}\}$
$RO_2H + RH \rightarrow RO^{\bullet} + {}^{\bullet}OH\}$ chain branching
$RO^{\bullet} + RH \rightarrow ROH + R^{\bullet}\}$
${}^{\bullet}OH + RH \rightarrow H_2O + R$
$RO_2H, RO_2^{\bullet}, R^{\bullet} \rightarrow$ scission and crosslinking
$2RO_2^{\bullet} \rightarrow RO_2R + O_2\}$ termination

Oxygen traps free radicals and drives reactions that are oxidative in nature. Some of these can be long lived which means over time they affect various properties in the irradiated material. These chemical reactions result in the formation of oxidative products which are created within the polymer and include the development of ketones, esters, alcohol derivatives, peroxide based species and carboxylic acids [40]. When peroxides degrade, chain branching is favoured and reactions result in further radical generation. In addition, the depth of oxidation in the polymer depends on the oxygen permeation rates and consumption rates per absorbed dose as well as the dose rate itself. These polymers have far less stability than the polymers irradiated in an inert atmosphere. Such post irradiation degradation schemes can render a polymer both oxidised and embrittled [38].

3. Materials and Methods

The polyurethane used in this study was a virgin Pellethane 2363 90A medical grade material manufactured by Lubrizol's. Tensile specimens and impact bar samples were manufactured using an Arburg injection moulding machine. The samples were irradiated between 0 and 200 kGy using a gamma irradiator (cobalt 60 source) and an electron beam irradiator (combined 10/12 MeV unit, 20 kW) at the Synergy Health facilities in Tullamore and Westport, Ireland [70, 71]. Information for the experimental procedures employed for some of the data presented are given in the respective references. In the case where unpublished results appear, melt flow index (MFI) values were measured according to ASTM standard D 1238 [73] by means of a CEAST Melt Flow Quick Index. MFI measurements were performed at a temperature of 224 °C, a 2.16 kg weight and a 2.00 mm diameter orifice during the space of 10 minutes. Dynamic viscosity measurements were carried out on the samples using an AR1000™ rheometer from TA instruments with a frequency range of 0.01

– 100 rad/s while using a temperature of 210 °C and a 1 % strain value. For tensile testing, an Instron 3365 machine was used with a crosshead speed of 500 mm/min and a 5 kN load cell. A Lloyd LRX Tensometer was used to calculate the flexural modulus of the impact bars while a type D shore durometer was used to calculate the hardness of the tensile specimens. Yellowness index was carried out according to ASTM D 1925 using a Lovibond RT600 sphere spectrophotometric colorimeter. The overall colour difference was established by implementing statistical analysis of the ΔE values for L, a, and b. ΔE was calculated by using the following formula [75]:

$$\Delta E = \sqrt{(\Delta L)^2 + (\Delta a)^2 + (\Delta b)^2} \tag{2}$$

ΔE is expressed as the difference between the control sample and the irradiated sample colour values.

4. Discussion and Background

Thermoplastic polyurethanes (TPU's) are segmented polymers comprising of alternating sequences of soft segments (polyether/polyester) and hard segments (diisocyanate and polyol) that form a unique microphase-separated structure [41]. This two-phase morphology provides the key to controlling performance of the final product and gives the manufacturer versatility in tuning properties as desired by varying the composition or content of one or the other phases [42]. Since the discovery of polyurethane by Otto Bayer and co-workers in 1937, polyurethanes have developed as a unique class of material and have found use in a broad range of applications [43]. They are used as coatings, adhesives, elastomers, foams, fibers, synthetic leathers and others, and are employed in the car industry, building and construction market, shoe manufacturing and medical technology [42]. Polyurethanes are widely used in the medical device industry to manufacture intracorporeal and extracorporeal devices such as angioplasty balloons, gastric balloons, roller pump, heart pump tubes, catheters and many more [44]. This is due to a combination of biocompatibility and excellent mechanical properties including high tensile strength, durability, flexibility and toughness the material has to offer [45, 46].

The urethane linkage is produced by the reaction of an isocyanate group of one reactant with the alcohol group of another component. By controlling variables such as the chemical composition, functionality, and the molecular weight of the different reactants, a wide class of materials with significantly varying properties can be obtained [47]. TPU's are linear polymers produced by the polymerisation reaction of three components:

- A diisocyanate (NCO—R—NCO)
- A short-chain diol, so-called chain extender (HO—R´—OH)
- A long-chain diol (OH OH)

The reaction for the formation of the urethane group is:

R—NCO + R´—OH R—NH—COO—R´

The structure of the linear polymeric chain of thermoplastic polyurethane is in two different types of block segments linked together by covalent links, producing a block copolymer. First of these is the hard segment which is formed by the reaction of the diisocyanate and the short-chain diol, which includes a high density of urethane groups of high polarity, thus rendering them rigid at room temperature. The second of these is the soft segment which is produced by the reaction of the diisocyanate and the long-chain diol. These have a low polarity as they have a very low density of urethane groups, and therefore, they are flexible at room temperature [48]. Depending on the diisocyanate used in the production of polyurethane, they can either be aromatic or aliphatic. Aromatic polyurethanes have better thermal resistance (in the absence of oxygen) than aliphatic polyurethanes. This is an important point, as the urethane group is the least thermal resistant group in the polyurethane structure [42, 49, 50]. Figure 7 shows a detailed diagram of these segments.

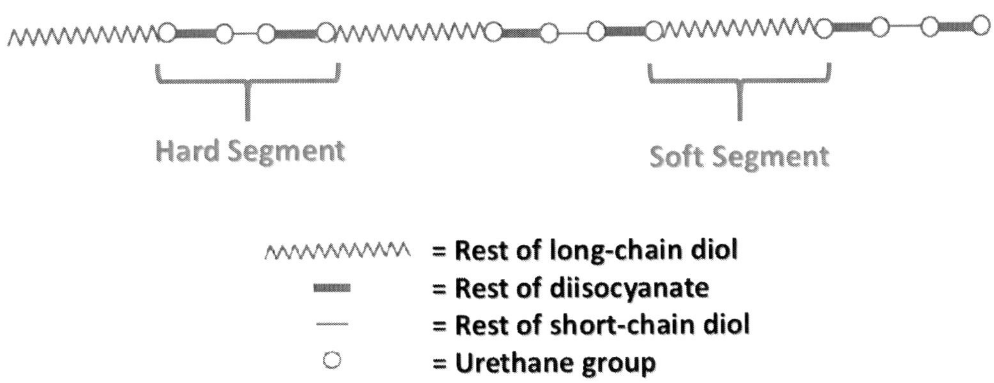

Figure 7. Structure of thermoplastic polyurethane.

Pellethane 2363 is a typical medical grade polyurethane which has been widely used as a biomaterial since its introduction in 1977 [51-53]. The chemical structure for this material is illustrated in Figure 8. Since the main application for this material is the manufacture of medical devices, it is essential that the end products are sterilised before use. However, such sterilisation processes like heat, dry steam and high energy irradiation can have detrimental effects on medical grade polymers such as extensive material degradation and plastic deformation [54-56]. During product development, the decision on a method of sterilisation is seen to be of little importance. Manufacturers of such medical devices often assume that existing sterilisation techniques are suitable and in most cases the key methods are satisfactory for the task. A lack of importance at this stage of the development process can suggest that it is a trivial decision [57]. Radiation sterilisation would be considered as an attractive form of sterilisation for polyurethane material as it offers excellent radiation resistance [58]. Aromatic polymers like polyurethanes are more resistant to high energy radiation than aliphatic polymers as they contain benzene rings [49, 59, 60]. The number of aromatic rings and the concentration of ester groups in the material determine its resistance to irradiation [6]. It has being previously reported that medical polyurethane products tolerate multiple exposures to gamma or beta irradiation. In consequence, they are commonly sterilised using these processes [54, 60, 61]. Nevertheless, previous work performed on biodegradable polyurethanes demonstrated a 20–50 % reduction in molecular weight and a 13–50 % decrease in tensile strength when exposed to a standard dose of 25 kGy used for sterilisation [44]. The extent of changes was strongly dependent on the chemical composition of the polyurethane [44, 49]. Radiation can either act chemically or physically leading to modifications of the structure or function of macromolecules which can result in oxidation, chain scission and crosslinking [4, 5]. Branching and crosslinking can be valuable to implant functionality in vivo and these are sometimes purposely introduced into the polymer to control degradation [44]. However, extremely high irradiation doses can have detrimental effects on such polymers and result in progressive degradation [62]. In terms of polyurethane, different types of degradation processes can evolve, such as the formation of quinoid structures and photo Fries rearrangement with C–N cleavage [63-66]. Photo-degradation is another problematic process which can occur upon the exposure of irradiation. These reactions occur in the backbone of the polyurethane structure and can lead to a loss of transmissivity, discolouration, crazing of the surface and embrittlement [67]. A previous study established that gamma radiation could lead to the release of MDA (methylene bis (p-phenyl aniline)) in MDI (methylene bis (p-phenyl isocyanate)) based polyurethanes, with correlation between the radiation levels and the release of MDA [68].

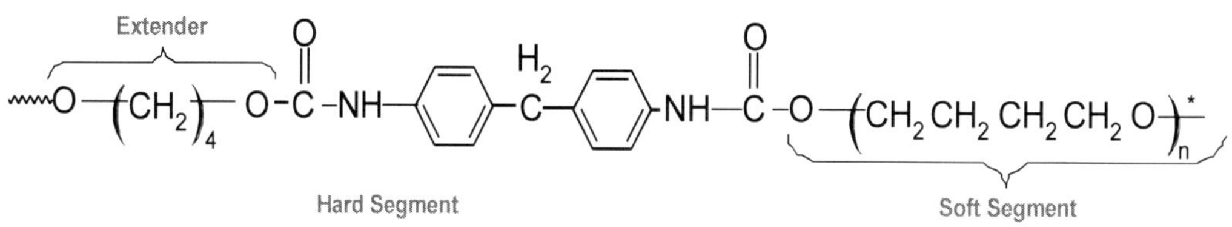

Figure 8. Chemical structure of Pellethane material [69].

4.1. Temperature of Polyurethane during the Irradiation Process

Temperature plays a significant role on the behaviour of polymers during an irradiation process. According to Kausch and Anjum [76], when the temperature of a given polymer is below the glass transition temperature during the irradiation process, a large quantity of stable radicals are generated and trapped in the glassy state. This in turn reduces the net crosslinking due to the immobility in the glassy state. However, when irradiation is performed on a given polymer above the glass transition temperature, the likelihood of crosslinking or scission is enhanced depending on the polymer chemistry [71]. Based on experimental data recorded by the author, there was a noticeable increase in the temperature of polyurethane (Pellethane 2363 90A) as the irradiation dose increased during the exposure to electron beam irradiation in an air atmosphere. Prior to irradiation treatment, each batch had a recorded temperature of 21 °C. Due to the limitations of the irreversible temperature labels, an exact temperature could not be recorded for temperatures below 37 °C. Figure 9 demonstrates the temperature increase throughout each of the dose ranges during the electron beam irradiation process. For the 5 and 25 kGy batches, the temperatures were ≥21 and <37 °C, however; above 25 kGy the temperatures increased with an increase in irradiation dose. A maximum temperature of 58 °C was achieved by the 200 kGy batch. These results suggest that temperature could have a substantial impact in promoting further degradation in the Pellethane material following electron beam treatment. As previously mentioned, a rise in 10 °C can double the reaction rates [77]. With regards to polyethylene, if this material is irradiated in liquid nitrogen temperatures, this will prevent the formation of allyl double bonds. Alternatively, if the polyethylene material is irradiated in the melt stage, the degree of crosslinking is significantly enhanced [76].

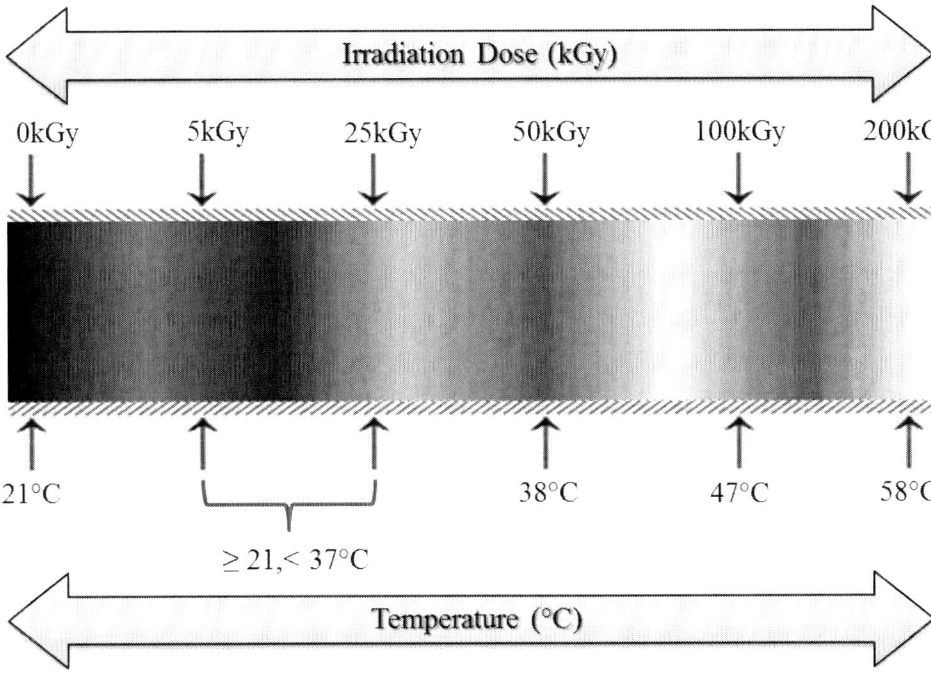

Figure 9. Temperature increase of the Pellethane material during the electron beam irradiation process.

4.2. Melt Flow Behaviour of Irradiated Polyurethane

The graph illustrated in Figure 10 represents the experimental outcome for each of the electron beam and gamma irradiated samples after the melt flow index (MFI) test. As the radiation dose incremented for the electron beam irradiated samples, the MFI decreased from 21g/10 mins at 0 kGy to 0.6 g/10 mins at 200 kGy. This signifies that the molecular weight drastically increased which in turn restricted the flow of the material. Such circumstances could be attributed to changes in the macro-molecules from radiation induced crosslinking or branching, which preferentially materialise in the amorphous phase. In contrast, the gamma irradiated samples at the lower dose rate (25 kGy) was identical to that of the electron beam samples as illustrated in Figure 10. However, at the higher dose rate (200 kGy), the gamma irradiated samples was less affected with an MFI value of 3.1g/10min in contrast to 0.6g/10mins for the electron beam irradiated samples. This was perhaps due to the gamma ray process generating less chain branches and for this reason the MFI instrument modified the material differently. Overall, from this experiment it was quite evident that chain branching/crosslinking was the predominant phenomenon which is in agreement to previous work carried out on the same material by Murray et al. [78].

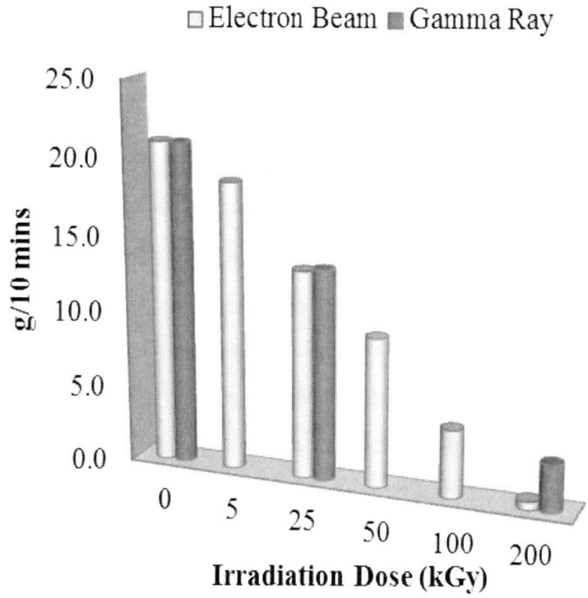

Irradiation Dose (kGy)	Standard Error	Standard Error
0	0.017	0.017
5	0.015	
25	0.013	0.011
50	0.060	
100	0.008	
200	0.002	0.005

Figure 10. Representation of the melt flow index of the non-irradiated and irradiated Pellethane samples.

Earlier work performed by Murray et al. [78] and Mrad et al. [79] have demonstrated that chain branching occurred in Pellethane 2363 material after the exposure of irradiation. Illustrated in Figure 11 is the chemical structure for Pellethane material which includes the proposed modified chemical structure as a result of radiation induced reactions. An alternative hypothesis was made by Shintani and Nakamura [80], for the crosslinking of gamma irradiated polyurethane which is exhibited in Figure 12. They stated that if crosslinking occurred at MDI-BU (methylene bis (*p*-phenylisocyanate)-1,4-Butanediol) hard segments as a minor reaction, secondary amino groups in the urethane bonds would be crosslinked to produce tertiary amino groups, therefore, reducing secondary amino groups in polyurethanes.

In addition to this, Shintani and Nakamura [80], proposed a mechanism for degradation as a result of gamma irradiated polyurethane. They emphasised that if cleavage and degradation transpired at the MDI-BU hard segments, then the reaction presented in Figure 13 would arise. Their work concluded that crosslinking and degradation mainly occurred at PTMG (polytetramethyleneglycol) soft segments by gamma irradiation.

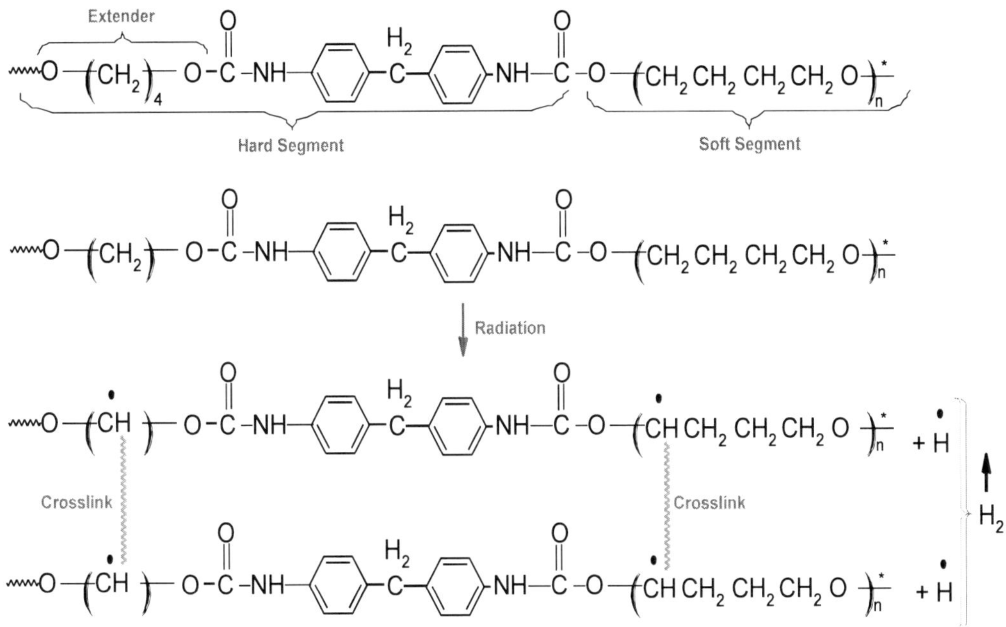

Figure 11. Proposed mechanism for crosslinking of irradiated Pellethane material [78].

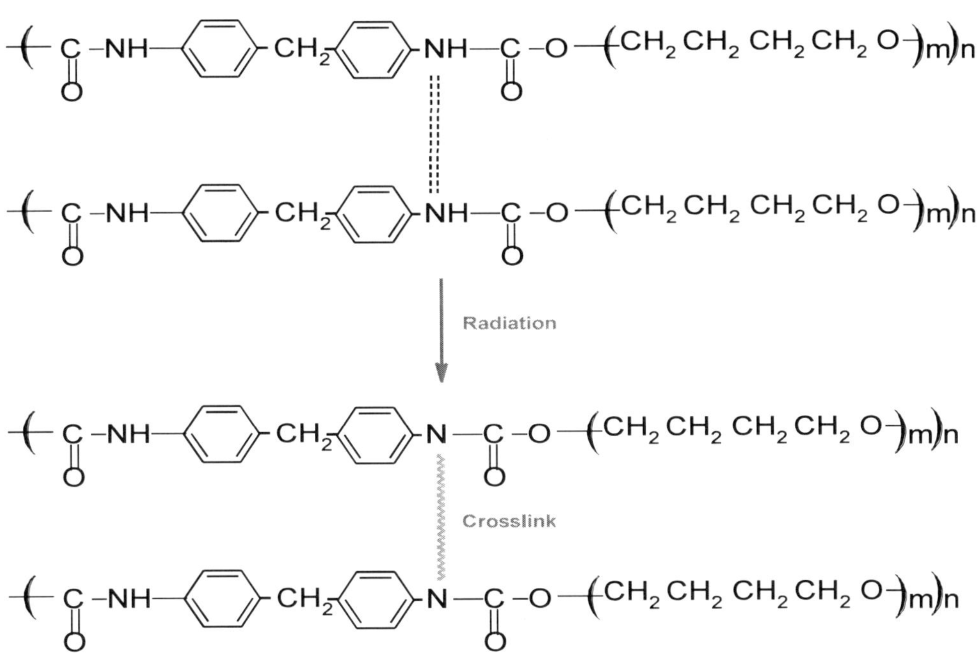

Figure 12. Proposed reaction scheme of crosslinking occurring at MDI-BU hard segments of non-chain-extended polyurethanes [80].

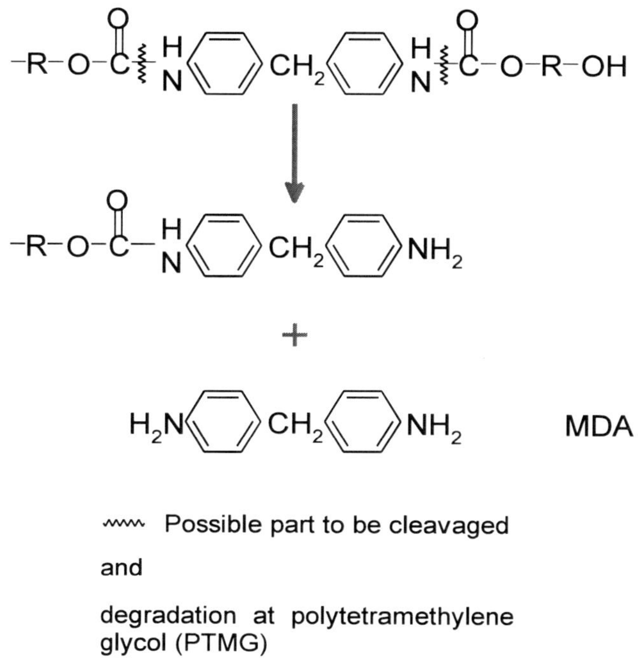

Figure 13. Proposed mechanism of degradation occurring at MDI-BU hard segments of chain-extended polyurethanes [80].

4.3. Rheology of Irradiated Polyurethane

Dynamic frequencies sweep experiments were carried out in the viscoelastic region (γ = 1.0 %) in order to study the microstructural changes and network formation of the irradiated samples while using the non-irradiated samples as the baseline. The plot of storage modulus (G'), loss modulus (G") versus angular frequency (ω) resulting from the dynamic frequency sweep scan measurements at 210°C are illustrated in Figures 14 and 15, respectively. Figure 14 demonstrates a comparative study of the electron beam and gamma irradiated Pellethane samples at 25 kGy. In the low frequency regime of the graph between 0.01 and 1 rad/s, it is evident that the G' increased marginally for the electron beam irradiated sample. However, the gamma irradiated sample had a significant increase in the G' indicating more structural alterations, which was perhaps related to the formation of chain branches. Furthermore, G" had a similar behaviour where the gamma irradiated sample promoted additional changes to occur in contrast to the electron beam irradiated sample and the non-irradiated sample. Based on Figure 14, it is evident that more modifications occurred at the low ω region as opposed to the high ω region of the graph. This suggests that the polymer chains were only somewhat affected by the frequencies in the high ω regime. Whereas in the low ω region there was an immense increase in the G' which could be attributed to the restriction on the movement of the polymeric chains by the development of a network structure [81].

Figure 15 represents the Pellethane samples irradiated with electron beam and gamma irradiation at 200 kGy. It is apparent from the graph that the 200 kGy irradiated samples had an increase in the elastic and viscous modulus in contrast to the 25 kGy samples. These modifications may have resulted from chain branching and gel formation caused by radiation exposure. Again, the gamma irradiated sample had a larger increase in the G' and G" in contrast to the electron beam irradiated sample. Overall, the results obtained confirm that the gamma ray process initiated more detrimental changes to occur in the Pellethane material.

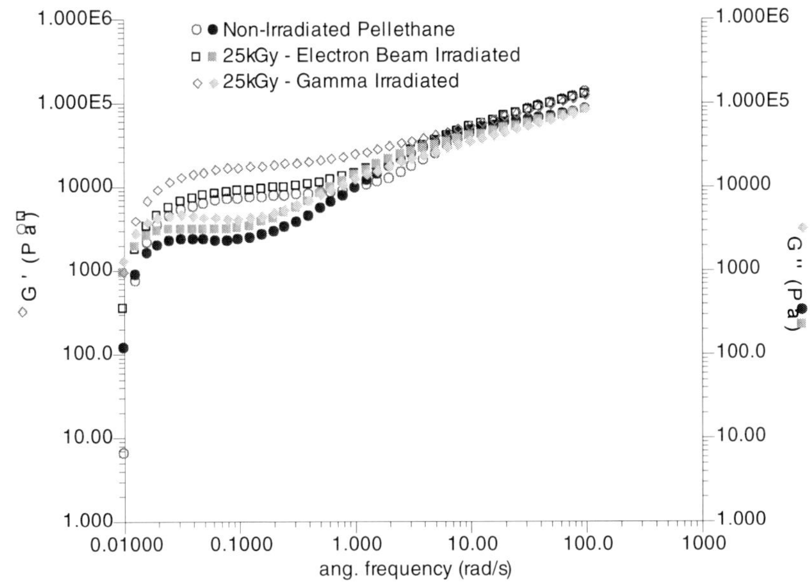

Figure 14. Comparative study of the non-irradiated and irradiated (gamma and electron beam) Pellethane material (G' and G" versus angular frequency) at 25 kGy.

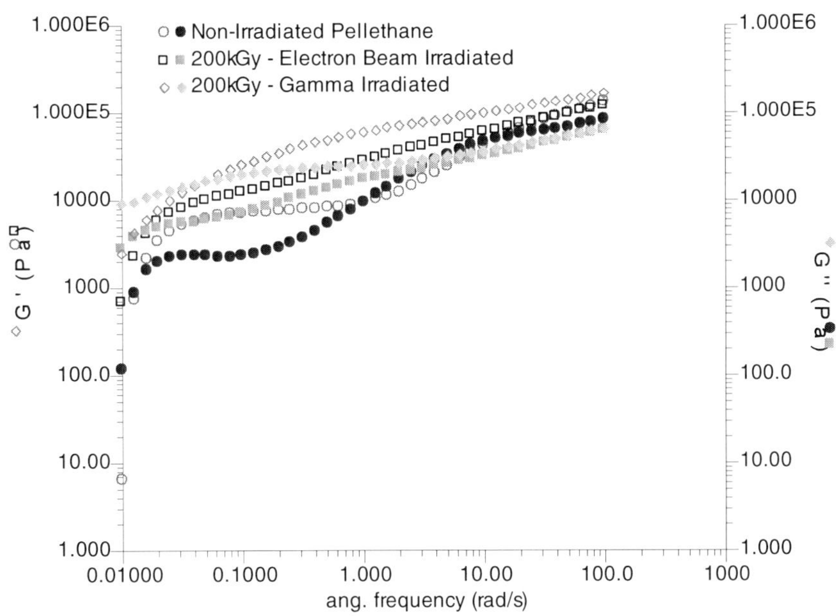

Figure 15. Comparative study of the non-irradiated and irradiated (gamma and electron beam) Pellethane material (G' and G" versus angular frequency) at 200 kGy.

4.4. Irradiation Effects on the Tensile Properties of Polyurethane

The tensile properties of thermoplastic elastomers generally depend on the size, shape and concentration of hard domains, intermolecular bonding within hard domains and the ability of soft segments to crystallise under strain [82]. Illustrated in Figure 16 is a graph representing the effects of electron beam and gamma irradiation on the tensile strength of the Pellethane material. While concentrating on the electron beam samples first, it is apparent that the tensile strength of the polymer decreased as the irradiation dose increased. Between 0 and 200 kGy, the tensile strength reduced from 54.68 MPa to 44.03 MPa which was perhaps attributed to the formation of chain scission and branching after the exposure of irradiation, which in turn affected the urethane bonds within the polymer. Gorna and Gogolewski [49] reported similar results in that the tensile strength decreased with radiation while utilising a gamma irradiator. However, a previous study has identified the occurrence of only chain scission in irradiated polyurethane [49, 79]. Based on the results from the MFI experiment and previous work performed on the same material [78], it can be suggested that chain branching/crosslinking was the dominating process. Branching can limit the packing of polymer chains and lead to the formation of free volume in the material [83-85]. This transformation can weaken the polymer structure which in turn reduces the tensile strength. Percentage elongation at break varied throughout the different dose ranges as presented in Figure 17. At 0 kGy, the percentage elongation at break was 846 % and this value increased to 871 % at 25 kGy followed by a decrease to 791 % at 200 kGy. At 25 and 100 kGy respectively, there was a slight increase. Overall, no significant changes were observed for the percentage elongation at break apart from the 200 kGy sample.

The gamma irradiated samples revealed to be more critical at lower irradiation doses and less critical at higher doses for the tensile strength and percentage elongation at break. Each of their values decreased from 0 to 25 kGy and then increased slightly as the irradiation dose increased to 200 kGy. Such changes could be attributed to chain branching and chain scission taking place simultaneously within the material. Otherwise it could be that one of the factors (chain branching or chain scission) was taking place at different rates throughout the different dose ranges. Overall, both irradiation processes affected the properties of the Pellethane material. Chain branching and/or chain scission were the main factors affecting such properties, where predominance was essentially dependant on the radiation type (electron beam or gamma) [86], the irradiation dose [87], the characteristics of the polyurethane [82] and the atmosphere during irradiation [88]. However, irradiation in an air/oxygen atmosphere generally leads to chain scission [88]. In the literature, Gorna and Gogolewski [49] reported gamma irradiated polyurethane leads to chain scission. In addition, Shintani et al. [89] reported similar results in that chain scission occurred when Pellethane 2363 was gamma irradiated. However, Mrad et al. [79] reported that chain branching was the predominant phenomenon for electron beam irradiated Pelletahne 2363 80A.

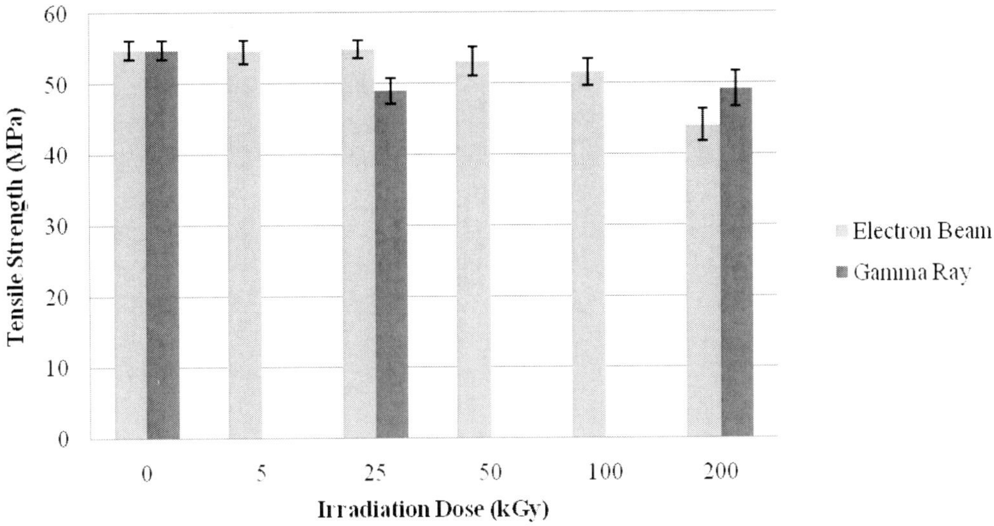

Figure 16. Graphical representation of the tensile strength versus irradiation dose of the Pellethane samples.

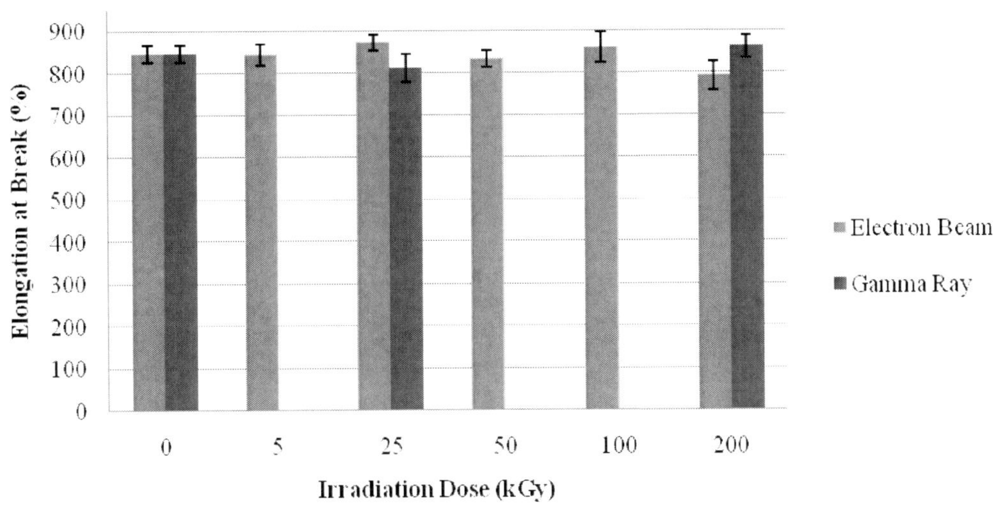

Figure 17. Graphical representation of the elongation at break versus irradiation dose of the Pellethane samples.

4.5. Stiffness Properties of Irradiated Polyurethane

Flexural modulus (stiffness) is an important physical property in many applications where it measures the behaviour of materials subjected to a simple beam loading. In this experiment the three point bend test produced a tensile stress in the convex side of the impact bar (specimen) and compression stress in the concave side. This developed an area of shear stress along the centre line of the ASTM sample. In this experiment the span to depth ratio was kept to 16. Other specimens/materials require S/d=32 to 64 to keep the shear stress low enough. The graph presented in Figure 18, illustrates the behaviour of the non-irradiated and irradiated Pellethane material after the three point bend test. An increase of 3.5 % in the flexural modulus between 0 and 5 kGy was observed for the electron beam irradiated samples, followed by a gradual decrease from 25-200 kGy with an overall percentage decrease of 8.0 %. Improvements to the flexural modulus are found to be in the lower irradiation dose range, whereby a maximum of 48.3 MPa was achieved at 5 kGy and 47.9 MPa at 25 kGy. At higher irradiation doses the flexural modulus of the material decreased with an increase in irradiation dose. The samples exposed to 200 kGy exhibited the greatest decrease, where it reached 42.9 MPa. An increase in the flexural modulus at the lower dose rates was perhaps attributed to radiation induced free radical reactions such as chain branching. Chain branching leads to the development of small branches along the polymeric chains within a material therefore, enhancing such material properties as stiffness. It is quite clear from the results obtained that the flexible properties of the Pellethane material were reduced at the higher dose ranges, but not to a large extent. Gamma irradiated samples resulted in a reduction in the flexural modulus according to an increase in the irradiation dose. The stiffness of the material reduced from 46.7 MPa at 0 kGy to 44.1 MPa at 25 kGy and then a further reduction was observed where it decreased to 34.9 MPa at 200 kGy. A reduction in the stiffness of a material is generally related to chain scission. Gorna and Gogolewski [49] reported similar results where the modulus decreased with radiation while utilising a gamma irradiator. These results indicate that branching was occurring between 5 and 25 kGy while chain scission and possibly branching was occurring simultaneously between 25 and 200 kGy for the electron beam irradiated samples. In terms of the gamma irradiated samples, chain scission appears to be the dominating factor subsequent to radiation exposure.

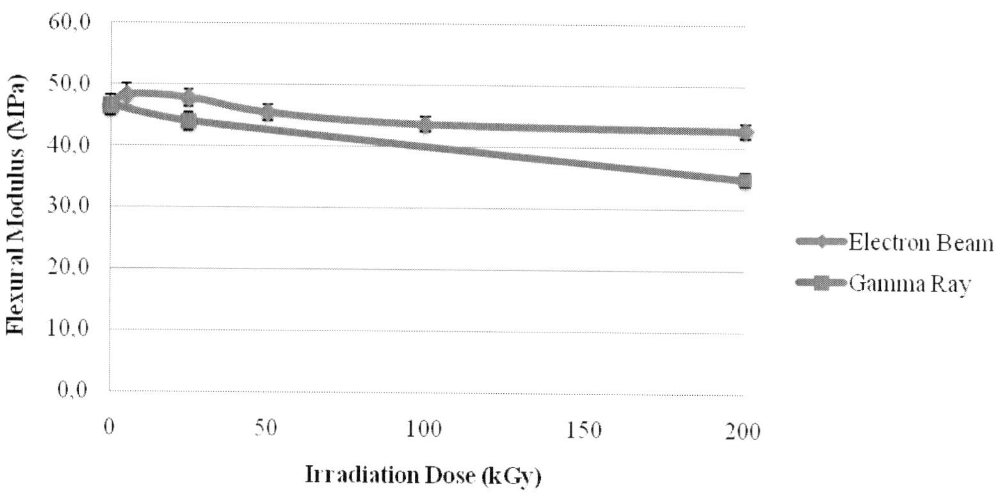

Figure 18. Effects of electron beam and gamma irradiation on the flexural modulus of the Pellethane material.

4.6. Effects of Irradiation on the Hardness of Polyurethane

A shore durometer was used in this experiment by the author to measure the resistance of the Pellethane material towards indentation and to provide an empirical hardness value. The Shore D scale was the preferred method in order to determine the hardness characteristics of the material while carrying it out according to ASTM D2240. Table 4 presents the data collected for the electron beam irradiated Pellethane samples. No significant alterations emerged, however it was evident that the hardness value increased slightly with radiation dose. As the irradiation dose incremented the hardness value increased in a linear fashion from 36.00 to 38.17 (D-scale). Therefore, the effects of post electron beam irradiation, increase the degree of chain branching/crosslinking and enhance molecular mass causing newly branched three dimensional networks. These results correlate with the findings from the MFI experiment on electron beam irradiated Pellethane, where both techniques signify the presence of branching/crosslinking.

Table 5 exhibits the results attained after the shore D test of the gamma irradiated samples. Here, only two irradiation doses were examined, the 25 and 200 kGy samples. It can be observed that the lower irradiation dose of 25 kGy affects the material considerably; more than the higher radiation dose of 200 kGy. The 25 kGy sample was modified to become softer in behaviour which could be attributed to the formation of chain scission after the exposure of radiation.

Table 4. Data collection for the Shore D hardness of electron beam irradiated Pellethane samples

Irradiation Dose (kGy)	Test 1	Test 2	Test 3	Test 4	Test 5	Test 6	Average	Standard Error
0	36.0	36.5	35.5	36.0	36.0	36.0	36.00	0.13
5	37.5	38.0	37.0	37.0	36.5	36.0	37.00	0.29
25	36.5	38.0	37.0	39.0	37.0	37.5	37.50	0.37
50	37.0	38.0	37.0	39.0	38.5	37.5	37.83	0.33
100	37.5	38.5	37.5	38.0	37.5	38.5	37.92	0.20
200	37.0	38.0	38.0	38.5	39.0	38.5	38.17	0.28

Table 5. Data collection for the Shore D hardness of gamma irradiated Pellethane samples

Irradiation Dose (kGy)	Test 1	Test 2	Test 3	Test 4	Test 5	Test 6	Average	Standard Error
0	36.0	36.5	35.5	36.0	36.0	36.0	36.00	0.13
25	36.5	35.0	34.0	36.0	35.0	36.0	35.42	0.37
200	34.5	37.0	36.0	35.0	36.0	38.0	36.08	0.52

4.7. Discolouration of Polyurethane Following Irradiation

Colorimetry was conducted on the non-irradiated and irradiated Pellethane samples in order to quantify the colour changes numerically. The graph plotted in Figure 20 illustrates the overall colour difference achieved for all batches subsequent to gamma irradiation and electron beam irradiation. The greatest overall change was exhibited by the gamma irradiated samples throughout all irradiation dose ranges and this can be seen visually in Figure 21. In the lower dose range of 25 kGy, a noteworthy change was observed as gamma irradiation discoloured the Pellethane material by more than double, in contrast to the electron beam irradiated samples. In the higher dose range (200 kGy), the difference in discolouration for both processes was more subtle. The electron beam irradiated samples obtained a value of 32.82 in comparison to 34.96 for the gamma irradiated samples. This evidently proves that the electron beam process leads to fewer alterations in terms of Pellethane discolouration. These modifications could be attributed to the oxidation of the central methylene group which in turn produces hydroperoxides. These hydroperoxides were perhaps the primary products of the reaction to form quinone imides (Figure 19) [63]. The urethane bridge oxidises due to the formation of these quinone imides which finally produces unsaturated chromophores leading to the yellowing of the Pellethane material. Chain scission may also have played an active role in this, whereby the formation of trapped free radicals could produce additional light absorbing structures like chromophores.

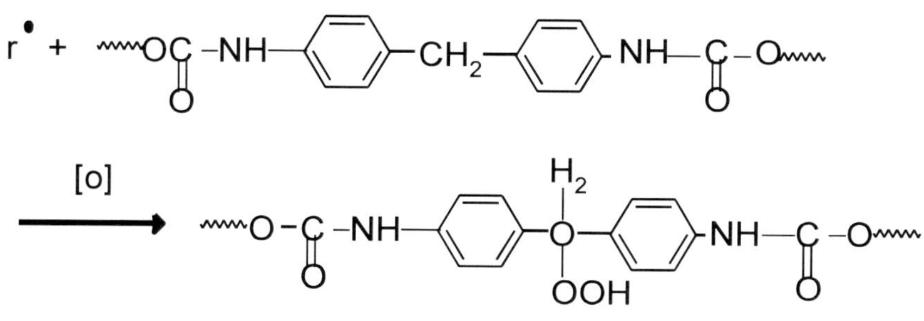

Figure 19. The illustration of the possible reaction process to Quinone-imides [63].

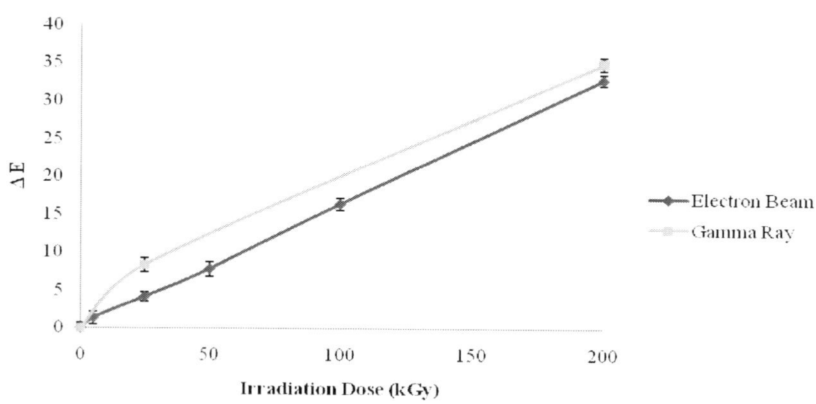

Figure 20. Graphical representation for the overall colour change (ΔE) of Pellethane as a result of electron beam and gamma irradiation.

Figure 21. Overall colour change of Pellethane as a result of electron beam and gamma irradiation by visual inspection.

CONCLUSION

In summary, irradiated Pellethane material demonstrated simultaneous reactions of branching and chain scission. However, a number of the experiments throughout this chapter and previous work performed by the author [78], illustrated chain branching was the predominant phenomena. This chapter highlighted that gamma radiation had a more undesirable effect on Pellethane material in contrast to electron beam radiation. Furthermore, a number of characterisation techniques provided evidence that the electron beam irradiated samples had fewer alterations at 25 kGy and more alterations at 200 kGy in contrast to the gamma irradiated samples. For example, the tensile strength behaviour was more affected by gamma irradiation at 25 kGy, whereas, electron beam irradiation resulted in more detrimental effects at 200 kGy. This outcome was perhaps attributed to the environmental conditions during the radiation processing of the samples. Irradiation of Pellethane material under inert gas atmosphere and/or lowering the radiation dose and/or lowering the temperature might be a potential method of preventing or at least diminishing the radiolysis of this material. Elevated temperatures would play a significant role in supporting such changes as this can result in the mobility of polymer chains, therefore, inducing a higher probability of radical chains modifications.

REFERENCES

[1] W. A. Rutala and D. J. Weber, "Infection control: the role of disinfection and sterilization," *Journal of Hospital Infection,* vol. 43, pp. S43-S55, 1999.

[2] N. A. Halls, "Gamma-irradiation Processing," in *Irradiation Effects on Polymers*, D. W. Clegg and A. A. Collyer, Eds., ed England: Elsevier Science Publishers Ltd, 1991, pp. 253-296.

[3] J. L. Williams, "Processing and mechanistic aspects of radiation sterilization of medical disposable plastic materials," in *Radiation processing of polymers*, A. Singh and J. Silverman, Eds., ed Munich, Vienna, New York, Barcelona: Hanser, 1991, pp. 255-270.

[4] A. L. Mazzu and C. P. Smith, "Determination of extractable methylene dianiline in thermoplastic polyurethanes by HPLC," *Journal of Biomedical Materials Research,* vol. 18, pp. 961-968, 1984.

[5] S. Lerouge, C. Guignot, M. Tabrizian, D. Ferrier, N. Yagoubi, and L. Yahia, "Plasma-based sterilization: effect on surface and bulk properties and hydrolytic stability of reprocessed polyurethane electrophysiology catheters," *Biomed Mater Res,* vol. 52, pp. 774-782, 2000.

[6] S. Pierpoint, J. Silverman, and M. Al-Sheikhly, "Effects of ionizing radiation on the aging of polyester based polyurethane binder," *Radiation Physics and Chemistry,* vol. 62, pp. 163-169, 2001.

[7] J. L. Bol. (2011, 5th August 2011). *X-Ray Sterilisation.* Available: http://www.med-techinnovation.com/Articles/articles/article/6.

[8] J. Meissner, M. Abs, M. R. Cleland, A. S. Herer, Y. Jongen, F. Kuntz, et al., "X-ray treatment at 5 MeV and above," *Radiation Physics and Chemistry,* vol. 57, pp. 647-651, 2000.

[9] M. Abs, Y. Jongen, E. Poncelet, and J. L. Bol, "The IBA rhodotron TT1000: a very high power E-beam accelerator," *Radiation Physics and Chemistry,* vol. 71, pp. 287-290, 2004.

[10] IBA. (2011, 10th August 2011). *High Power X-ray Technology.* Available: http://www.iba-sterilization.com/e-beam-x-ray/x-ray.

[11] Y. Jongen, "Industrial Applications of Accelerators: Traditional and New," ed. Washington: IBA, 2006.

[12] C. G. Dunn, W. L. Campbell, H. Fram, and A. Hutchins, "Biochemical and *photo-chemical effects* of *high* energy, electrostatically produced roentgen rays and cathode rays," *Applied Physics,* vol. 19, pp. 605-615, 1948.

[13] IAEA, *Trends in radiation sterilization of health care products.* Vienna: International Atomic Energy Agency, 2008.

[14] J. Cuda, "Industrial gamma processing overview," in *AECL Gamma Processing Seminar*, Ottawa, 1983.

[15] F. M. Fraser, "Cobalt-60 availability for radiation processing," in *17th Japan Conference on Radiation and Radioisotopes*, Tokyo, 1985.

[16] E.-S. A. Hegazy, "Gamma Radiation Sterilization Service Centre in a Developing Country: Egypt," in *Trends in radiation sterilization of health care products*, ed Vienna: International Atomic Energy Agency, 2008, pp. 81-89.

[17] Sterigenics. (2004, 11th August 2011). *Gamma Radiation.* Available: http://www.sterigenics.com/services/medical_sterilization/contract_sterilization/gamma_radiation/Gamma%20Alt%2017_0.pdf.

[18] P. Dethier, J. L. Bol, and B. Mullier. (2010, 8th August 2011). *X-ray Sterilisation: The Technology of the Future.* Available: http://www.emdt.co.uk/article/x-ray-sterilisation-technology-future.

[19] V. J. Boaler, "Electron Beam Processing," in *Irradiation Effects on Polymers*, D. W. Clegg and A. A. Collyer, Eds., ed Essex: Elsevier Science Publishers, 1991, pp. 297-317.

[20] Sterigenics. (2004, 9th May 2011). *Sterilization Alternatives: Electron Beam Radiation*. Available: http://www.sterigenics.com/services/medical_sterilization/contract_sterilization/electron_beam_radiation/sterilization_alternatives__electron_beam_radiation.pdf.

[21] A. G. Chmielewski, T. Sadat, and Z. Zimek, "Electron Accelerators for Radiation Sterilisation," in *Trends in radiation sterilization of health care products*, ed Vienna: IAEA, 2008, pp. 27-47.

[22] K. S. S. Sarma, "Development of a Family of Low, Medium and High Energy Electron Beam Accelerators," presented at the Emerging applications of radiation processing, Vienna, 2004.

[23] S. Tagawa, M. Kashiwagi, T. Kamada, M. Sekiguchi, K. Hosobuchi, H. Tominaga, et al., "Economic Scale of Utilization of Radiation (I): Industry. Comparison between Japan and the USA," presented at the Nuclear Science and Technology, Japan, 2002.

[24] A. J. Berejka, "Advances in self-shielded accelerators," presented at the Emerging applications of radiation processing, Vienna, 2004.

[25] A. J. Berejka and M. R. Cleland. (2011, 9th May 2011). *Industrial Radiation Processing With Electron Beams and X-rays*. Available: http://www.cirms.org/pdf/Industrial%20Radiation%20Processing-May%202011-Revision%206.pdf.

[26] M. Silindir and A. Y. Ozer, "Sterilization Methods and the Comparison of E-Beam Sterilization with Gamma Radiation Sterilization," *Pharm. Sci,* vol. 34, pp. 43–53, 2009.

[27] IAEA, *Use of mathematical modelling in electron beam processing: a guidebook*. Vienna: IAEA, 2010.

[28] I-Ax Technologies Inc. (1999, 25th August 2011). *Electron Beam Accelerators*. Available: http://www.iaxtech.com/more.html.

[29] S. K. Datta, T. K. Chaki, and A. K. Bhowmick, "Electron Beam Processing of Polymers," in *Advanced Polymer Processing Operations*, N. P. Cheremisinoff, Ed., ed New Jersey: Noyes Publications, 1998, pp. 157-186.

[30] S. A. Mine and O. Yekta, "Sterilization Methods and the Comparison of E-Beam Sterilization with Gamma Radiation Sterilization," *Pharm. Sci,* vol. 34, pp. 43–53, 2009.

[31] O. James, "Electron Beam Compatible Polymers," in *ANTEC, 62nd Annual Technical Conference of Society of Plastics Engineers*, Chicago, IL, 2004.

[32] J. G. Drobny, *Radiation Technology for Polymers*. New York: CRC Press, 2003.

[33] A. Ahmad, D. H. Mohd, and I. Abdullah, "Electron Beam Cross-linking of NR/LLDPE Blends," *Iranian Polymer Journal,* vol. 14, pp. 505-510, 2005.

[34] W. Rogers, "Radiation Sterilisation," in *Sterilisation of Polymer Healthcare Products*. vol. 1, R. T. Limited, Ed., 1 ed Shropshire: Rapra Technology Limited, 2005, pp. 147 - 204.

[35] Z. P. Zagorski, "Thermal and electrostatic aspects of radiation processing of polymers," in *Radiation Processing of Polymers*, ed Munich: Hanser, 1992, pp. 271-287.

[36] C. Thevy Ratnam and K. Zaman, "Effect of electron beam irradiation on surgical rubber gloves," *Nuclear Instruments and Methods in Physics Research Section B: Beam Interactions with Materials and Atoms,* vol. 140, pp. 171-177, 1998.

[37] M. Burton, "Discuss," *Faraday Society,* pp. 317-318, 1952.

[38] L. A. Pruitt, "The Effects of Radiation on the Structural and Mechanical Properties of Medical Polymers " *Advances in Polymer Science,* vol. 162, pp. 63-93, 2003.
[39] H. Wasserman and R. Murray, "Singlet oxygen," ed. New York: Academic Press, 1979.
[40] R. L. Clough, K. T. Gillen, and C. A. Quintana, "Heterogeneous oxidative degradation in irradiation polymers," *Journal of Polymer Science,* vol. 23, pp. 359 -377, 1985.
[41] S. A. Madbouly and J. U. Otaigbe, "Recent advances in synthesis, characterization and rheological properties of polyurethanes and POSS/polyurethane nanocomposites dispersions and films," *Progress in Polymer Science,* vol. 34, pp. 1283-1332, 2009.
[42] M. Rogulska, A. Kultys, and W. Podkoscielny, "Studies on thermoplastic polyurethanes based on new diphenylethane-derivative diols. II. Synthesis and characterization of segmented polyurethanes from HDI and MDI," *European Polymer Journal,* vol. 43, pp. 1402-1414, 2007.
[43] J. P. Theron, "Development of crosslinkable, thermoplastic polyurethanes for cardiovascular prostheses," Doctor of Philosophy (Chemical Engineering), Process Engineering, Stellenbosch, Stellenbosch, South Africa, 2006.
[44] K. Gorna and S. Gogolewski, "Molecular stability, mechanical properties, surface characteristics and sterility of biodegradable polyurethanes treated with low-temperature plasma," *Polymer Degradation and Stability,* vol. 79, pp. 475-485, 2003.
[45] S. Abd El All, "Investigations on the electrical and structural properties of electron-beam irradiated polyurethane," *J. Phys. D: Appl. Phys.,* vol. 40, pp. 6014-6019, 2007.
[46] D. J. Wilson, N. P. Rhodes, and R. L. Williams, "Surface modification of a segmented polyetherurethane using a low-powered gas plasma and its influence on the activation of the coagulation system," *Biomaterials,* vol. 24, pp. 5069-5081, 2003.
[47] B. D. Kaushiva, "Structure-Property Relationships of Flexible Polyurethane Foams," Doctor of Philosophy in Chemical Engineering Enginering, Virginia Polytechnic Institute and State University, Blacksburg, Virginia, 1999.
[48] Merquinsa. (2003, 12th September 2011). *What's A Polyurethane.* Available: http://www.merquinsa.com/whats/whatsaPU.pdf.
[49] K. Gorna and S. Gogolewski, "The effect of gamma radiation on molecular stability and mechanical properties of biodegradable polyurethanes for medical applications," *Polymer Degradation and Stability,* vol. 79, pp. 465-474, 2003.
[50] K.-Y. Chen, J.-F. Kuo, and C.-Y. Chen, "Synthesis, characterization and platelet adhesion studies of novel ion-containing aliphatic polyurethanes," *Biomaterials,* vol. 21, pp. 161-171, 1// 2000.
[51] M. Szycher, "Biostability of polyurethane elastomer: A critical review," in *Blood Compatible Materials and Devices*, C. P. Sharma and M. Szycher, Eds., ed Lancaster, PA: Technomic Publishing, 1990, pp. 33-85.
[52] A. Coury, K. Stokes, P. Cahalan, and P. Slaikeu, "Biostability considerations for implantable polyurethanes," *Life Support Systems,* vol. 5, pp. 25-39, 1987.
[53] K. Stokes, A. Coury, and P. Urbanski, "Auto-oxidative degradation of implanted polyether polyurethane devices," *Biomaterials Applications,* vol. 1, pp. 411-448, 1987.
[54] I. P. Matthews, C. Gibson, and A. H. Samuel, "Sterilisation of implantable devices," *Clinical Materials,* vol. 15, pp. 191-215, 1994.
[55] S. Gogolewski and P. Mainil-Varlet, "Effect of thermal treatment on sterility, molecular and mechanical properties of various polylactides: 2. Poly(l/d-lactide) and poly(l/dl-lactide)," *Biomaterials,* vol. 18, pp. 251-255, 1997.

[56] W. E. Skiens and J. L. Williams, *Ionizing Radiation's Effects on Selected Biomedical Polymers*. Lancaster, PA: Technomic, 1983.

[57] A. Simmons, J. Hyvarinen, and L. Poole-Warren, "The effect of sterilisation on a poly(dimethylsiloxane)/poly(hexamethylene oxide) mixed macrodiol-based polyurethane elastomer," *Biomaterials,* vol. 27, pp. 4484-4497, 2006.

[58] N. Lamba, K. A. Woodhouse, S. L. Cooper, and M. D. Lelah, *Polyurethanes in Biomedical Applications*. Boca Raton, FL: CRC Press, 1998.

[59] S. D. Bruck, "Sterilization problems of synthetic biocompatible materials," *Journal of Biomedical Materials Research,* vol. 5, pp. 139-158, 1971.

[60] M. Goldman and L. Pruitt, "Comparison of the effects of gamma radiation and low temperature hydrogen peroxide gas plasma sterilization on the molecular structure, fatigue resistance, and wear behavior of UHMWPE," *Journal of Biomedical Materials Research,* vol. 40, pp. 378-384, 1998.

[61] M. Goldman, R. Gronsky, G. G. Long, and L. Pruitt, "The effects of hydrogen peroxide and sterilization on the structure of ultra high molecular weight polyethylene," *Polymer Degradation and Stability,* vol. 62, pp. 97-104, 1998.

[62] M. d. Lelah and S. L. Cooper, *Polyurethanes in Medicine*. Boca Ration, FL: CRC Press, 1986.

[63] C. Wilhelm, A. Rivaton, and J.-L. Gardette, "Infrared analysis of the photochemical behaviour of segmented polyurethanes: 3. Aromatic diisocyanate based polymers," *Polymer,* vol. 39, pp. 1223-1232, 1998.

[64] B. Ravat, R. Gschwind, M. Grivet, E. Duverger, A. Chambaudet, and L. Makovicka, "Electron irradiation of polyurethane: Some FTIR results and a comparison with a EGS4 simulation," *Nuclear Instruments and Methods in Physics Research Section B: Beam Interactions with Materials and Atoms,* vol. 160, pp. 499-504, 2000.

[65] B. Ravat, M. Grivet, Y. Grohens, and A. Chambaudet, "Electron irradiation of polyesterurethane: study of chemical and structural modifications using FTIR, UV spectroscopy and GPC," *Radiation Measurements,* vol. 34, pp. 31-36, 2001.

[66] B. Ravat, M. Grivet, and A. Chambaudet, "Evolution of the degradation and oxidation of polyurethane versus the electron irradiation parameters: Fluence, flux and temperature," *Nuclear Instruments and Methods in Physics Research Section B: Beam Interactions with Materials and Atoms,* vol. 179, pp. 243-248, 2001.

[67] K. M. Zia, I. A. Bhatti, M. Barikani, M. Zuber, and M. A. Sheikh, "Thermo-mechanical characteristics of UV-irradiated polyurethane elastomers extended with α, ω-alkane diols," *Nuclear Instruments and Methods in Physics Research Section B: Beam Interactions with Materials and Atoms,* vol. 267, pp. 1811-1816, 2009.

[68] H. Shintani and A. Nakamura, "Formation of 4,4′-methylenedianiline in polyurethane potting materials by either γ-ray or autoclave sterilization," *Journal of Biomedical Materials Research,* vol. 25, pp. 1275-1286, 1991.

[69] Elastogran. (2005, 12th September 2011). *Thermoplastic Polyurethane Elastomers (TPU)*.Available: http://www.polyurethanes.basf.de/pu/solutions/de/function/conversions:/publish/content/group/Arbeitsgebiete_und_Produkte/Thermoplastische_Spezialelastomere/elastollan_material_uk.pdf.

[70] K. A. Murray, J. E. Kennedy, B. McEvoy, O. Vrain, D. Ryan, R. Cowman, et al., "Effects of gamma ray and electron beam irradiation on the mechanical, thermal,

structural and physicochemical properties of poly (ether-block-amide) thermoplastic elastomers," *Journal of the Mechanical Behavior of Biomedical Materials,* vol. 17, pp. 252-268, 1// 2013.

[71] K. A. Murray, J. E. Kennedy, B. McEvoy, O. Vrain, D. Ryan, R. Cowman, et al., "The effects of high energy electron beam irradiation in air on accelerated aging and on the structure property relationships of low density polyethylene," *Nuclear Instruments and Methods in Physics Research Section B: Beam Interactions with Materials and Atoms,* 2013.

[72] K. A. Murray, J. E. Kennedy, B. McEvoy, O. Vrain, D. Ryan, and C. L. Higginbotham, "The effects of high energy electron beam irradiation on the thermal and structural properties of low density polyethylene," *Radiation Physics and Chemistry,* vol. 81, pp. 962-966, 8// 2012.

[73] ASTM D638-03, "Standard test method for tensile properties of plastics," ed. West Conshohocken, PA: ASTM International, 1994.

[74] K. A. Murray, J. E. Kennedy, B. McEvoy, O. Vrain, D. Ryan, R. Cowman, et al., "The effects of high energy electron beam irradiation in air on accelerated aging and on the structure property relationships of low density polyethylene," *Nuclear Instruments and Methods in Physics Research Section B: Beam Interactions with Materials and Atoms,* vol. 297, pp. 64-74, 2/15/ 2013.

[75] R. Perera, C. Albano, J. González, P. Silva, and M. Ichazo, "The effect of gamma radiation on the properties of polypropylene blends with styrene-butadiene-styrene copolymers," *Polymer Degradation and Stability,* vol. 85, pp. 741-750, 2004.

[76] H. Kausch and N. Anjum, *Radiation Effects on Polymers for Biological Use*: Springer, 2003.

[77] A. J. Berejka and M. R. Cleland. (2011, Industrial Radiation Processing With Electron Beams and X-rays. 34. Available: http://www.cirms.org/pdf/Industrial%20 Radiation%20Processing-May%202011-Revision%206.pdf.

[78] K. A. Murray, J. E. Kennedy, B. McEvoy, O. Vrain, D. Ryan, R. Cowman, et al., "The influence of electron beam irradiation conducted in air on the thermal, chemical, structural and surface properties of medical grade polyurethane," *European Polymer Journal.*

[79] O. Mrad, J. Saunier, C. Aymes Chodur, V. Rosilio, F. Agnely, P. Aubert, et al., "A comparison of plasma and electron beam-sterilization of PU catheters," *Radiation Physics and Chemistry,* vol. 79, pp. 93-103, 2010.

[80] H. Shintani and A. Nakamura, "Mechanism of degradation and crosslinking of polyurethane when irradiated by gamma-rays," *Journal of Applied Polymer Science,* vol. 42, pp. 1979-1987, 1991.

[81] H. Liu, Z. Fang, M. Peng, L. Shen, and Y. Wang, "The effects of irradiation cross-linking on the thermal degradation and flame-retardant properties of the HDPE/EVA/magnesium hydroxide composites," *Radiation Physics and Chemistry,* vol. 78, pp. 922-926, 2009.

[82] G. A. Abraham, P. M. Frontini, and T. R. Cuadrado, "Physical and mechanical behavior of sterilized biomedical segmented polyurethanes," *Journal of Applied Polymer Science,* vol. 65, pp. 1193-1203, 1997.

[83] D. Bamford, G. Dlubek, T. Lüpke, D. Kilburn, J. Stejny, T. J. Menke, et al., "Free Volume, Glass Transition and Degree of Branching in Ethylene/α-Olefin Copolymers:

Positron Lifetime, Differential Scanning Calorimetry, Wide-Angle X-Ray Scattering, and Density Studies," *Macromolecular Chemistry and Physics,* vol. 207, pp. 492-502, 2006.

[84] J. E. Puskas, Y. Chen, K. Kulbaba, G. Kaszas, and A. Soleymannezhad, "Effect of the molecular weight and architecture on the size and glass transition of arborescent polyisobutylenes," *Journal of Polymer Science Part A: Polymer Chemistry,* vol. 44, pp. 1770-1776, 2006.

[85] D. Kritskaya, S. Kurmaz, and I. Kochneva, "Glass transition temperature and architecture of branched poly(methyl methacrylates)," *Polymer Science Series A,* vol. 49, pp. 1120-1128, 2007.

[86] L. Woo and C. L. Sandford, "Comparison of electron beam irradiation with gamma processing for medical packaging materials," *Radiation Physics and Chemistry,* vol. 63, pp. 845-850, 2002.

[87] C. Guignot, N. Betz, B. Legendre, A. Le Moel, and N. Yagoubi, "Degradation of segmented poly(etherurethane) Tecoflex® induced by electron beam irradiation: Characterization and evaluation," *Nuclear Instruments and Methods in Physics Research Section B: Beam Interactions with Materials and Atoms,* vol. 185, pp. 100-107, 2001.

[88] A. Chapiro, "General consideration of the radiation chemistry of polymers," *Nuclear Instruments and Methods in Physics Research Section B: Beam Interactions with Materials and Atoms,* vol. 105, pp. 5-7, 1995.

[89] H. Shintani, H. Kikuchi, and A. Nakamura, "Effects of gamma-ray irradiation on the change of characteristics of polyurethane," *Polymer Degradation and Stability,* vol. 32, pp. 17-30, 1991.

In: Biological Responses, Monitoring and Protection ... ISBN: 978-1-63463-852-4
Editor: K. P. Mishra

Chapter 11

TREATMENT OF ACUTE RADIATION SYNDROMES WITH TRANSPLANTATION OF BONE MARROW-DERIVED MESENCHYMAL STROMAL CELLS

Claudia Lange, Rudolph Reimer, Jozef Zustin and Bärbel Brunswig-Spickenheier
Clinic for Stem Cell Transplantation, Dept. Cell and Gene Therapy, University Medical Center Hamburg-Eppendorf, Hamburg, Germany

ABSTRACT

Survival in radiation induced bone marrow (BM) failure depends on the dose of radiation received and the intensity of supportive care which can protect from otherwise lethal infection and give surviving stem cells a chance to expand. In cases of limited hematopoietic recovery potential, hematopoietic stem cell (HSC) transplantation is the only curative treatment option. The acute radiation syndrome (ARS) similarly to myelosuppression after total body irradiation (TBI) is not restricted to the BM function but can lead to development of multiple organ dysfunctions. Damage of the whole organism is related to a systemic inflammatory response caused by a release of inflammatory cytokines. Management of ARS therefore relies on tissue damage repair processes that might be supported by therapies directed at mitigation of inflammation.

We have shown previously the regenerative and immunomodulating effects of BM-derived mesenchymal stromal cells (MSC) e.g. in acute kidney injury. Furthermore, MSC augment hematopoietic recovery after chemo- or radiotherapy transplanted in conjunction with hematopoietic stem cells and are effective in the treatment of steroid resistant graft versus host disease. Therefore we tested the potency of MSC as a treatment option after severe radiation exposure.

As a first step, human MSC were subjected to in vitro differentiation into hematopoietic and endothelial lineages. In a second approach, mouse MSC were isolated from BM of mice and i.v. transplanted into lethally irradiated recipients. For long-term observation, leukocyte and thrombocyte counts were determined regularly. After 7 months, all available tissues were investigated for the presence of donor cells. Short-term detection of donor cells was done using quantitative PCR. BM from HSC- and MSC-transplanted animals at day 21 was investigated for differential gene expression.

In vitro we showed aquisition of hematopoietic and endothelial characteristics, albeit to a rather low degree. In vivo, the transplantation of mouse MSC led to normalization of white blood cells within 4 weeks. Quantitative evaluation of donor cells in recipients revealed a fast disappearance of donor MSC. Although donor MSC did not home to the BM, the gene expression profile in BM changed significantly. Genes involved in cell cycle, oxidative stress protection, anti-inflammation and detoxification showed upregulation in the BM of MSC-transplanted animals whereas genes responsive for inflammation, protein degradation, adhesion/matrix formation, lipid synthesis and lymphoid development were downregulated, emphasizing the beneficial role of MSC in endogenous hematopoietic reconstitution.

In conclusion, we showed a complete reconstitution of the blood system in lethally irradiated mice through transplantation of MSC even without HSC support. Because no homing of donor MSC to the BM was evident we concluded that MSC help to recover from lethal myelosuppression and salvage endogenous HSC. This conclusion was supported by chimerism analysis showing no differentiation of donor cells into HSC in vivo. Instead, the milieu in the BM is changed towards a survival-promoting environment counteracting deleterious effects provoked by irradiation. Potential mechanisms of this highly effective trophic mechanism of MSC and its perspectives will be discussed.

INTRODUCTION

The management of patients suffering from acute radiation syndromes (ARS) still remains a major challenge. Survival of radiation-induced bone marrow failure depends on the dose of radiation received and the intensity of supportive care which can protect from otherwise lethal infection and give surviving stem cells a chance to expand. To provide the best possible care for radiation accident victims in acts of terrorism or catastrophic incidences, medical countermeasures need to be made within the first few days for optimal efficacy [1]. High dose radiation induced tissue damage results in little probability for autologous recovery. Combined approaches including presenting symptoms, biomarkers and physical dosimetry are employed to categorize affected individuals for best medical countermeasures [2].

Overall measures include supportive care, treatment with growth factors within the first two weeks after radiation exposure, or hematopoietic stem cell transplantation (HSCT). Since radiation effects on blood stem cells occur at doses generally lower than on other critical organs, the rapidly emerging changes in the peripheral blood cell lineages dictate the treatment options. In fact, total body irradiation (TBI) at doses more than 7-8 Gy in human corresponds to medullar eradication. Under this threshold spontaneous recovery from residual hematopoietic stem and progenitor cells may be expected within 30–50 days but going through cytopenic phases of granulocytic, megakaryocytic and erythrocytic lineages. HSCT should be considered if the victim's hematopoietic stem cell (HSC) pool is essentially irreversibly damaged. Interestingly, even after TBI, intrinsically radioresistant stem cells have been detected in distinct bone marrow (BM) areas comprising a residual hematopoietic stem and progenitor cell pool [3]. ARS does not only imply damage to the bone marrow. In a dose-dependent matter, it can also emerge as gastrointestinal and cerebrovascular syndromes leading to development of multiple organ dysfunctions [3]. Damage to the whole organism is related to a systemic inflammatory response. Different target organs are affected due to activation of the innate immune system, resulting in a significant release of inflammatory cytokines [4]. The pathophysiology appears comparable to that of acute graft-versus-host disease (GvHD) following allogeneic stem cell transplantation where a similar "cytokine storm" has been observed [5]. Long-term effects of ionizing radiation have been well documented in atomic bomb survivors in whom persistent signs of inflammation, e.g. increased plasma levels of tumor necrosis factor-α (TNF-α), interferon-β (IFN-β), interleukin-6 (IL-6), and C-reactive protein, have been reported [6]. Additionally, oxidative stress after high dose ionizing radiation has been involved in delayed morbidity [3]. Management of ARS therefore relies on tissue damage repair processes that might be supported by therapies directed at mitigation of inflammation [3].

Efforts to improve outcome for affected individuals focus on the stem cell niche. Therefore, visionary therapies should augment niche activity to accelerate hematopoietic recovery in vivo. Several studies have demonstrated that BM osteoblasts regulate the HSC pool size in vivo via the Jagged1-Notch signaling pathway [1]. For example, parathyroid hormone receptor activation can increase the number of osteoblastic cells, resulting in Notch1-mediated expansion of HSC [7]. One integrative part of the BM stroma are the mesenchymal stromal cells (MSC), also described as osteoblastic progenitors [8]. MSC are multipotential nonhematopoietic progenitor cells capable of differentiating into multiple lineages of the mesenchyme. In the bone marrow, the niche stromal cells surround the HSC and their progeny. The niche provides a sheltering microenvironment that supports the maintenance and self-renewal of HSC by shielding them from differentiation and apoptotic stimuli that would otherwise challenge stem cell reserves. Moreover, the niche also controls the proliferation and differentiation of HSC and the release of mature progeny into the vascular system. The regulation of HSC quiescence, through the maintenance of HSC in the G0 phase of the cell cycle in the endosteal niche, and the control of HSC proliferation, differentiation and recruitment in the vascular niche can be ascribed to bone-marrow stromal cells [9]. Thus, the physiological function of MSC is not the direct replacement of mesenchymal tissues such as bone, but their primary and most important function is to inhibit immunosurveillance and to establish a protective and regenerative microenvironment for HSC.

Clinically, MSC have been proven to intervene with acute organ impairments. Cotransplanted with HSC, MSC augment hematopoietic recovery after chemo- or radiotherapy significantly decreasing the time to full hematopoietic and particularly platelet reconstitution [10]. Additionally, there is a large body of evidence for MSC effectiveness in the treatment of steroid resistant GvHD without any side effects even when obtained from BM of third-party donors [11, 12]. No HLA-matching is needed between donor and recipient because MSC have been shown to be hypoimmunogenic and are not recognized by the recipients immune system even after repeated injections [11, 12]. Finally, MSC secrete a plethora of bioactive molecules [13, 14]. Among these, several essential hematopoietic growth factors including IL-6, IL-11, leukemia inhibitory factor (LIF), stem cell factor (SCF) and Flt3 ligand are produced but also factors with immunomodulatory effects, e.g. transforming growth factor-β1 (TGF-β1), prostaglandin E2, indoleamine 2,3-dioxygenase, and others [14, 15]. Additionally, vascular endothelial growth factor (VEGF) secreted by MSC in large amounts might interfere with an early apoptotic cell death after irradiation [3]. Therefore, MSC might be a good candidate for the modulation of hematopoietic niche activity.

In summary, MSC have emerged as a promising therapeutic tool for tissue regeneration and repair. Further clinical interest has been raised by the observation that MSC are immunoprivileged and might be used from unreleated, i.e. allogeneic donors [9, 15]. Altogether, we assumed that MSC with their comprehensive trophic potential could serve as a readily available treatment option after severe radiation exposure.

METHODS AND RESULTS

In Vitro Differentiation of Human MSC (hMSC)

As first experiments, we investigated the capability of BM-derived MSC to differentiate into progenitors for hematopoietic (HSC) and endothelial cells (EC). Human MSC were thoroughly characterized according to the ISCT (International Society for Cellular Therapy) criteria [16], comprising flow cytometry and differentiation capability into three mesodermal lineages [17]. To avoid any contamination of MSC with HSC, cloned cells were used exclusively. Cloned human MSC were subjected to differentiation into (i) hematopoietic cells using serum-containing or serum-depleted growth conditions and (ii) endothelial cells (for methodical details see ref. 18). Fibroblastoid MSC (Figure 1a) formed blast-like cells with noticeably decreased diameter from originally 28.9 ± 6.6 to 15.7 ± 3.5 μm during the differentiation into hematopoietic (Figure 1b) and endothelial (Figure 1c) lineages. The in vitro conditions led to cluster formation appearing as an in vitro equivalent of stromal structures from which differentiation proceeded. Hematopoetically differentiated cells changed their gene expression towards a typical profile of progenitor (CD117, CD133, CD45) and mature (CD14, CD16, glycophorin GlyA, CD31, podoplanin PDPN) hematopoietic cells (Figure 2a). Interestingly, the erythropoietin receptor (EPOR) was upregulated in almost all clones and under all conditions suggesting a definite role for EPO in proliferation and differentiation of mesodermal progenitors. Additionally, a variety of transcription factors responsible for erythropoiesis (SCL/tal1), erythro-megakaryopoiesis (GATA1, GATA2), lymphopoiesis (GATA3), and myelopoiesis (NOTCH1, RUNX1) were upregulated upon serum-containing differentiation. As SCL and RUNX1 are transcription factors essential for HSC formation by instructing lineage specification [19, 20] we hypothesized an efficient induction of this differentiation pathway in MSC. Using immunofluorescence, antigen-positive cells with small round or polymorphic nuclei were detected showing the expression of hematopoietic progenitor and mature antigen expression (not shown, refer to ref. 18), albeit to a rather low degree. In parallel, the same cells were able to aquire endothelial morphology and expressed endothelial genes upon cultivation with endothelial promoting factors (Figure 2b). At the protein level, single double positive cells for CD31/vWF (von Willebrand factor) and VEGFR-2/CD34 were detected [18].

Hematopoietic and endothelial progenitors share expression of a number of genes, including VEGFR-2, CD34, SCL, GATA2, RUNX1, and CD31, suggesting that investigated hMSC possess hemangioblastic capacity in vitro, and the extrinsic factors used might play a decisive role in the differentiation process as lineage-instructive regulators.

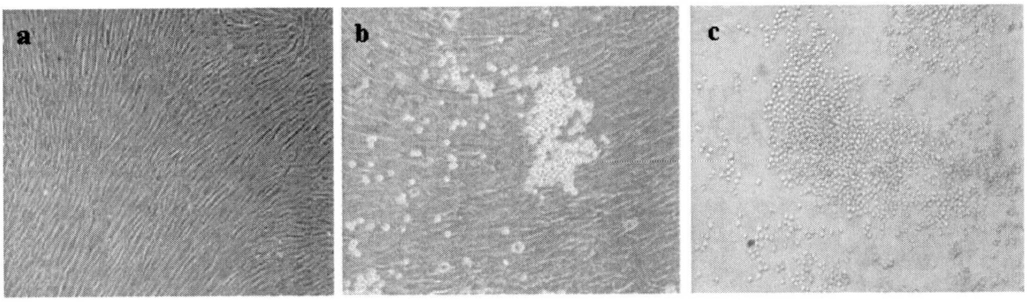

Figure 1. Human MSC display a fibroblastoid morphology during in vitro expansion but form blast-like cells after induction of differentiation. One clonal hMSC culture is shown during expansion (a), differentiation into hematopoietic (b) or endothelial (c) cells.

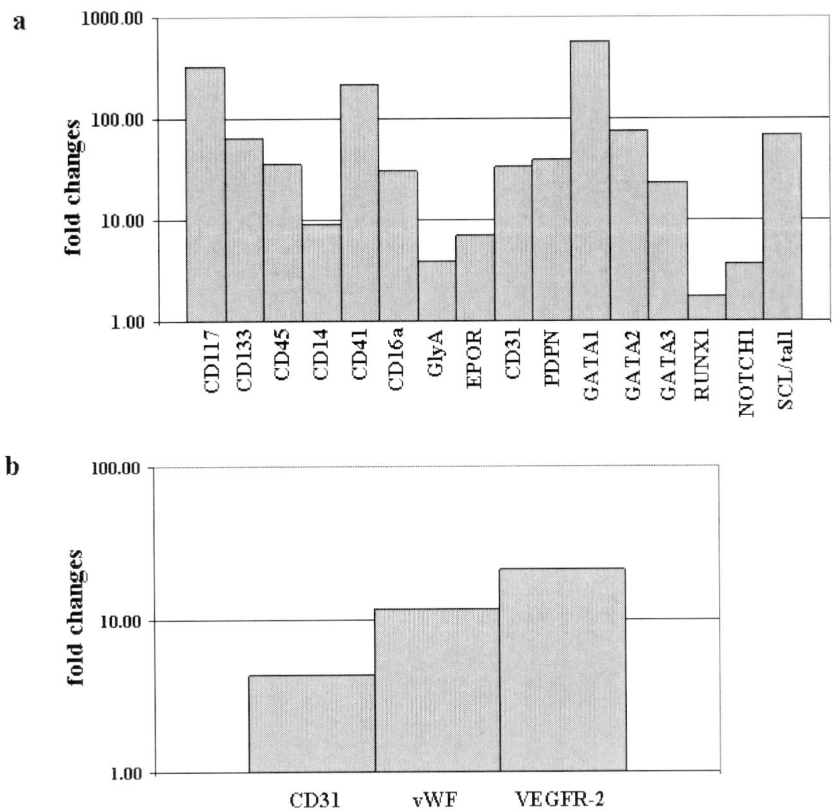

Figure 2. Human MSC significantly upregulate expressions of hematopoietic and endothelial genes after induction of differentiation. Shown are the fold changes of gene expressions of indicated hematopoietic (a) and endothelial (b) genes after differentiation compared to undifferentiated hMSC.

Most potent differentiation was achieved in cultures where the majority of hMSC adopted stroma function instructing a minor part to differentiate.

From the in vitro results we concluded that MSC might reconstitute the hematopoietic system. Hypothetically, one pluripotent stem cell would suffice to rescue lethally irradiated hosts, however in reality approx. 6 cells are needed [21], i.e. six pluripotent MSC with the respective potential might suffice.

MSC Promote Hematopoietic Recovery after Lethal Irradiation

To test the in vivo capability of MSC to replenish the hematopoiesis after eradication, lethally irradiated female recipients of the C57Bl/6J-CD45.1 strain were i.v. transplanted with 10^6 eGFP-marked male bulk-culture mouse MSC (mMSC). Leukocyte and thrombocyte recovery was similar to recipients transplanted with HSC (Figure 3) reaching normalization of white blood cells after 4 weeks. Seven months later, the recipients were hematologically well with a normal distribution of peripheral cell populations (Table 1). Similar experiments were carried out with clonal mMSC showing one clone (IXH8) with superior survival promoting properties (Table 2). Impressively, clone IXH8 was morphologically different from all other cultures showing increased CD34 and CD45 but no CD105 expression (Table 2). Downregulated CD105 as a marker of proliferation and adhesion [22] might increase survival of recipients by lowering lung embolization. All other characteristics corresponded to the ISCT criteria [16], questioning the relevance of CD105 expression for MSC characterization. Accordingly, over-expression of CD105 in cultured ECs induced a marked increase in protein levels of inflammatory eNOS [23], suggesting an anti-inflammatory action of CD105-negative mMSC in our model.

Transplanted Donor Cells Are Detectable Short- but not Long-Term

To detect donor chimerism in recipients, we stained recipient peripheral blood (PB), BM and thymus cells with CD45.2 antibodies and carried out flow cytometry.

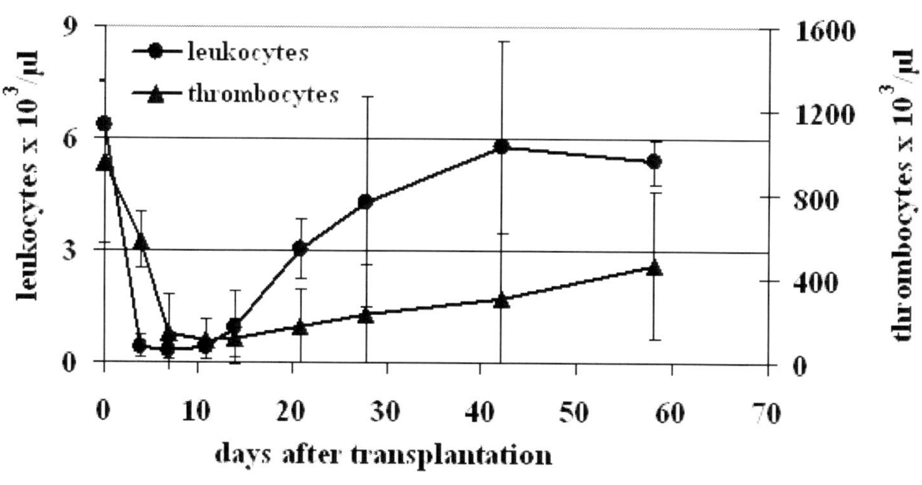

Figure 3. Mouse MSC rescue mice after total body irradiation. Transplantation of bulk mMSC led to a normalization of the peripheral white blood cell count within 4 weeks. Thrombocyte recovery needed approx. 8 weeks for normalization.

Table 1. Peripheral blood cell populations in mMSC transplanted animals

Lymphocytes	neutrophils	monocytes	eosinophils
72 % ± 3	21 % ± 3	5 % ± 2	2 % ± 1

Shown is the distribution of white blood cells 5 months after bulk mMSC transplantation Estimated using Pappenheim-stained blood smears.

Table 2. Phenotypical characterization of mMSC and recipients' survival rates after transplantation

	CD34	CD45	CD59	CD90	CD105	CD117	Sca-1	survival at 7 months [%]
bulk	1.6	0.5	95.4	0.5	85.9	0.9	96.7	19/28 [67.9]
IXH8	***9.8***	***4.1***	97.4	2.7	***1.6***	1.5	99.2	15/17 [88.2]
IVH7	1.2	1.3	54.7	0.5	94.1	2.8	81.9	2/12 [16.7]
IXC2	0.9	2.2	79.6	1.2	94.0	1.5	90.2	3/10 [30]
VIIIE7	1.2	1.1	71.1	2.0	93.1	1.5	96.4	4/10 [40]
VF10	2.2	2.2	45.9	0.7	74.0	3.4	77.9	3/10 [30]
radiation control	nd	nd	nd	nd	nd	nd	nd	0/15 [0]

Cultures of eGFP-transduced bulk and cloned mMSC after extended expansion were positive for CD59, CD105 and Sca-1 but negative for the hematopoietic markers CD34, CD45, CD117 and for CD90 by flow cytometry. Clone IXH8 was different from all other cultures in its expression of CD34/CD45 and negativity of CD105 (shown in bold italic). Transplantation with this clone resulted in the highest survival rate of the irradiated recipients, suggesting elevated CD34 and CD45 and no CD105 expressions might be a prerequisite of the high rescue capability. nd, not done.

Interestingly, no CD45.2-positive cells were found at any time point, thus not showing regeneration through donor cells.

Y-chromosome-based chimerism analysis in female recipients using specific Y-chromosome primers for quantitative PCR could not detect donor cells in any investigated tissue including PB and BM (not shown), although animals survived long-term. Spectral karyotyping of clonal mMSC detected the loss of the Y-chromosome (Figure 4), whereas bulk cultures of passage13 were still Y-positive (not shown). Thus, mMSC might lose sex-specific chromosomes.

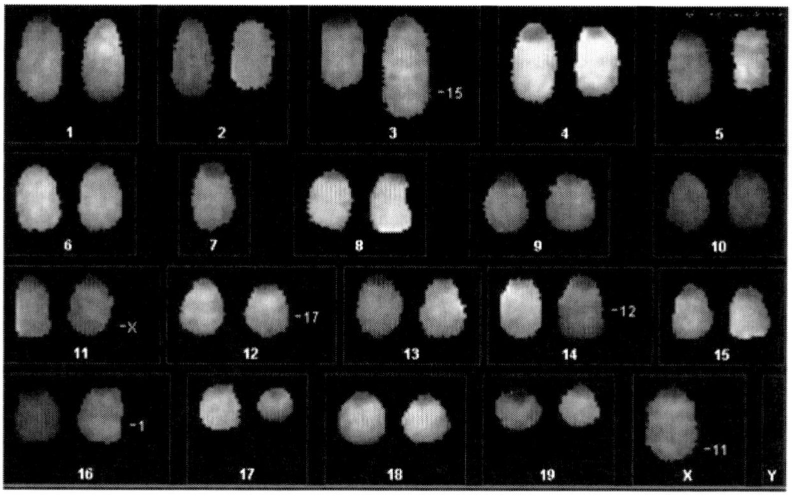

Figure 4. Spectral karyotyping of mMSC. Shown is the SKY analysis of clone IXH8. SKY analysis of a representative diploid metaphase revealed the loss of the Y-chromosome and this has been observed in all metaphases analyzed.

Next, we used eGFP-specific primers for quantitative PCR-based donor cell detection. Primers for stably integrated eGFP-sequences however also failed to detect any donor cells and no eGFP-positive cells were found in PB, BM or thymus by flow cytometry (not shown). Although we cannot completely rule out single donor cells below the detection limit, hematopoietic recovery in recipients is unlikely due to replenishment with donor cells. This conclusion contradicts earlier results of hematopoietic recovery after myeloablative TBI with blood-derived mMSC [24, 25] showing donor characteristics in blood and BM. One fundamental difference between both cell sources is the immortalization, potentially altering BM seeding capability of MSC. Therefore, our results support the concept of impaired transplantability of expanded MSC [26-28] but also challenge the hypothesis of high plasticity of MSC [29].

The distribution kinetics of eGFP+ donor cells after i.v. transplantation substantiated the fast disappearance from PB, reaching ca. 2% after 8 hours and no cells at d10 (Figure 5a). In contrast, mMSC trapped in lungs quickly (Figure 5b), however without long-term residence and embolization as shown by lack of donor signals after d10. Accordingly, no donor cells were evident in the spleen, liver, BM (Figure 5b), aorta, kidney, intestine, fat, thymus or lymph nodes (not shown). Although we did not find donor derived MSC in the BM, the morphology of this organ was preserved by MSC transplantation showing a normal distribution of different compartments (Figure 6). Without MSC, adipocytes seem to take over within short time, thus destroying the marrow structure.

MSC Change the BM Gene Expression

While donor mMSC did not home to the BM, we observed a long term survival and assumed an influence of MSC on the BM function. Therefore we carried out microarray analysis of bone marrow from MSC transplanted animals and compared the gene expressions to that of HSC-transplanted animals and aged matched controls [18].

a

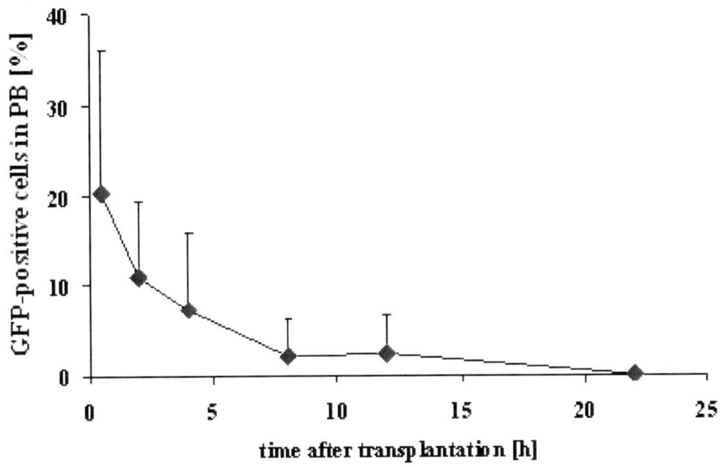

b

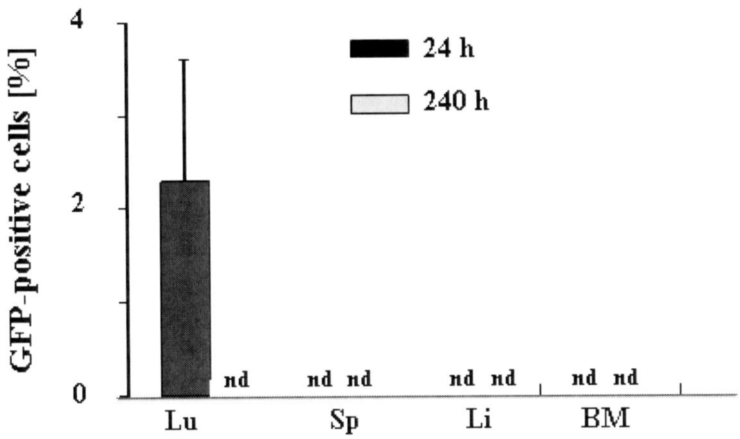

Figure 5. Donor mMSC are not detectable long-term. (a) Tracking of eGFP-labeled clonal IXH8 donor mMSC after transplantation revealed a fast decrease in peripheral blood (PB). Within 8 hours, approx. 2% were quantified in PB and none after 10 days (n = 8 for each time point). (b) mMSC accumulated in lungs (Lu) within 24 h and disappeared within 10 days (240 h). Spleen (Sp), liver (Li) and BM were negative at d1 and d10. nd, not detected.

The gene expression profile in BM changed significantly, clustering as separate group compared to untreated BM or HSC-transplanted mice. Validation of selected genes with high variance proved the beneficial role of MSC in endogenous hematopoietic reconstitution [18]. MSC in a complex mechanism upregulated protection from oxidative stress, cell cycle, anti-inflammatory and detoxification processes (e.g. gene expressions of BRPK, Cdkn1a, Thbs2, Gstm5) whereas lymphoid development, proinflammation, protein degradation and adhesion/matrix formation for improved cell motility (e.g. gene expressions of Vpreb1, Rag2, Klk6, Klk1b5, Uchl1, Sykb, Gpam, Col5a3, Emid1) were downregulated [18].

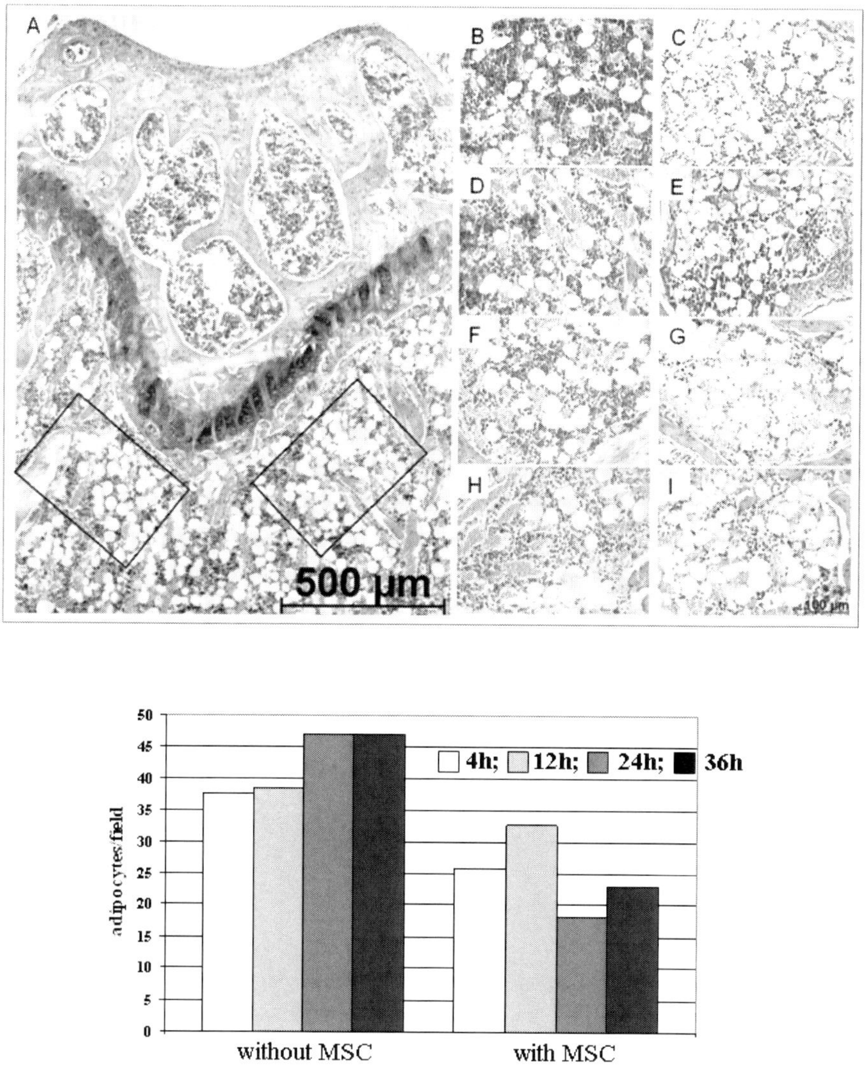

Figure 6. Histomorphology of BM with and without MSC transplantation. Paraffin embedded long bones from MSC-transplanted or control animals were cut and the number of adipocytes counted in 2 designated areas (A) per bone from mice with MSC transplantation after 4 (B), 12 (D), 24 (F) and 36 (H) hours or without MSC transplantation (C, E, G, and I respectively). The lower figure shows the number of adipocytes at each time point.

Summarising the microarray expression data we showed the upregulation of genes which are beneficial to the reconstitution of the BM whereas genes with supposed radiation-related BM deterioration were downregulated (Figure 7).

Potential Paracrine Mechanism of MSC

The questions how MSC, which were trapped in the lung, transfer their protective effects to the BM remains an open question. Recently, we could show that injection of MSC-derived microvesicles resulted in a similar protection as MSC transplantation (Figure 8).

Microvesicles are ultra-small bilipidlayer particles of 30-1000 nm (including exosomes) known to shuttle proteins, lipids, mRNA and microRNA [30]. Each of the components could participate in the radiation protection when transferred to the bone marrow. Interestingly, the reconstitution of platelets occurred at a faster time frame compared to MSC. Further work has to reveal the precise mechanism how microvesicles confer the radiation protection.

DISCUSSION

We provide evidence that donor MSC do not reconstitute the hematopoietic system but salvage endogenous irradiation-surviving HSC. Irradiation produces excessive inflammatory responses [31], contributing to HSC death if untreated. Among other organs, the lung is heavily involved in irradiation damage and might retard MSC.

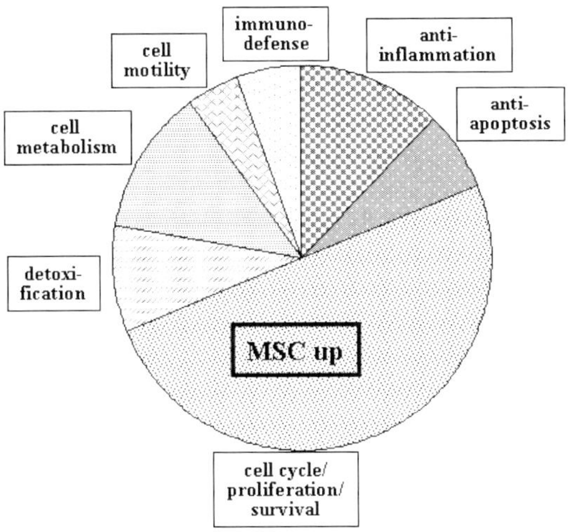

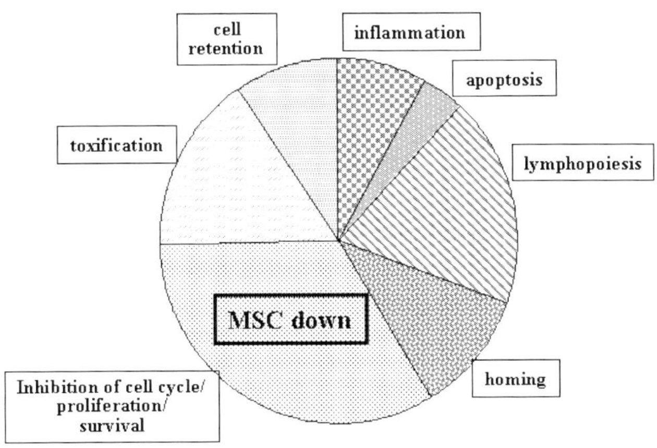

Figure 7. MSC transplantation into lethally irradiated animals changes the gene expression in the bone marrow. Gene expression data were generated using microarray analysis and significantly regulated genes clustered into functional groups. Shown are functional gene groups upregulated (MSC up) or downregulated (MSC down) in MSC-transplanted animals.

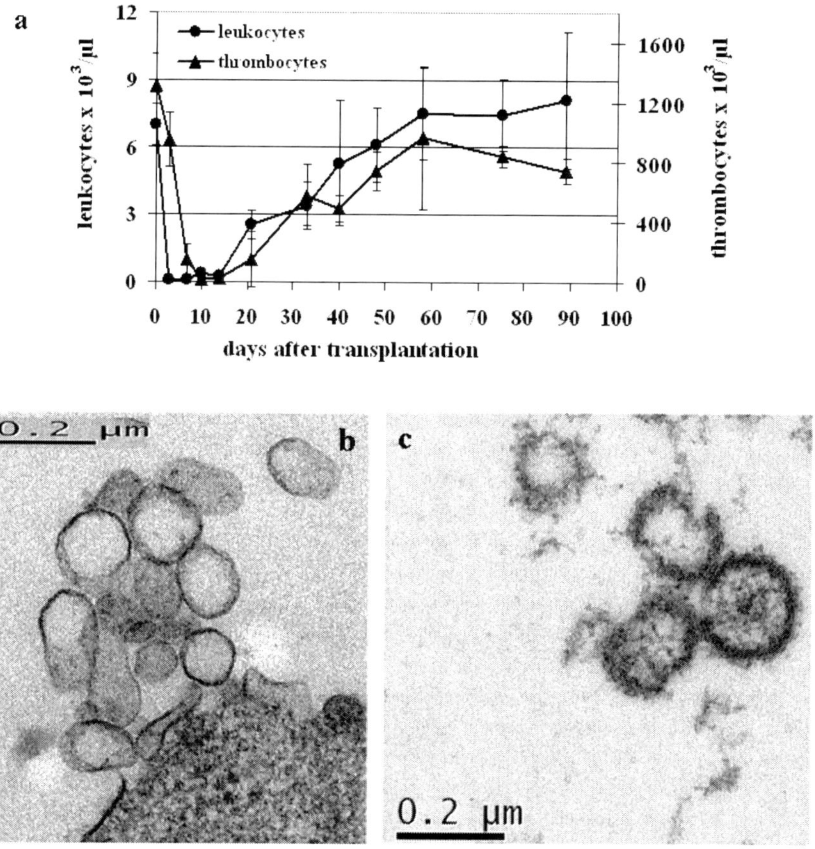

Figure 8. Transplantation of mMSC-derived microvesicles rescues lethally irradiated animals. Leukocyte counts after mMSC-derived microvesicle injection normalized with similar kinetics as with mMSC, whereas thrombocyte counts showed a much faster normalization (a). Electron microscopy of microvesicles released from mMSC (b) and purified by ultracentrifugation (c).

MSC interfere with inflammation by changing the expression profile not only in the lungs where they dock [32] but also in BM. To do so, MSC not necessarily need to home to BM but might globally change gene expression. These mechanisms have been described for MSC influencing myocardial infarction [32]. Similarly to mMSC, injected hMSC trapped in lungs and changed gene expression producing antiapoptotic TSG6 which enhances myocardial repair without significant engraftment. Differentiation-independent paracrine MSC-effect also ameliorate kidney injury [33-36].

What we have to expect from MSC as a therapeutic tool? Secretion of a broad range of bioactive molecules is now believed to be the main mechanism by which MSC achieve their therapeutic effect [37]. MSC secrete bioactive molecules that (a) inhibit apoptosis and limit the extent of damage or injury; (b) inhibit fibrosis or scarring at sites of injury; (c) protect the microvasculature and stimulate angiogenesis to improve perfusion; and (d) stimulate the mitogenesis of tissue-intrinsic progenitor cells, as has been shown for cardiac-, neural- and kidney-specific stem cells [9, 33-36]. In parallel, in ARS we showed that MSC boosted anti-inflammation, anti-apoptosis, detoxification, cell cycle and anti-oxidative stress control whereas proinflammatory, ECM formation and adhesion properties were decreased. In general, systemic improvements counteracting deleterious effects of myelosuppression are the result of MSC injection [18].

Thus, transplanted MSC might export their inherent trophic effect to unorthodox sites [38]. Our results present another piece of evidence for this highly effective paracrine mechanism working e.g. in BM, suggesting MSC-infusion as efficient and immediate treatment option after irradiation injuries.

TRANSLATION OF PRECLINICAL OBSERVATIONS INTO CLINICAL TRIALS

Bone marrow derived MSC can be readily generated from a small-volume of bone-marrow aspirate, which is obtained via a minimally invasive and safe procedure. The cells can subsequently be expanded in culture on a large scale, enabling the production of a standardized cell product that is suitable for use in various potential therapeutic applications [39-41]. The treatment of severe GvHD, a life-threatening complication after allogeneic, hemopoietic stem cells transplantation, was the first application in which the efficacy of MSC was demonstrated and no serious adverse events even after multiple infusions administered to several hundred patients have been recorded [11]. Importantly, MSC might be safely used without human leukocyte antigen matching and autologous or allogeneic MSC have been given without any side effects [11, 42]. New protocols substituting animal sera in expansion media have been established and shown to produce MSC with similar or better immunomodulating activities [39].

Despite the limitations in our existing knowledge, the capacity of MSC to exert their therapeutic effect through bystander mechanisms also might indicate that persistent engraftment at the site of damage is not a mandatory prerequisite for having a protective and curative effect on injured tissues, and possibly on local progenitors during acute stress conditions.

Importantly, the very brief residence time of MSC in the injured kidney and/or the entire organism might critically contribute to the safety of this cell-based therapy. The essential next step will be the conduct of clinical trials in which the definite efficacy of this MSC-based or even microvesicle therapy is tested.

ACKNOWLEDGMENT

This work was supported by the Federal Ministry of Education and Research, Germany, grant number 13N8904 and by the "Deutsche José Carreras Leukämie-Stiftung e.V.", grant number DJCLS R 12/30.

CONFLICT OF INTERESTS

There are no commercial associations that might create a conflict of interest in connection with this book chapter.

References

[1] Fliedner T. M., Chao N. J., Bader J. L., Boettger A., Case C. Jr. et al. (2009) Stem cells, multiorgan failure in radiation emergency medical preparedness: a U.S./European Consultation Workshop. *Stem Cells*, 27: 1205-1211.

[2] Drouet M., Hérodin F. (2010) Radiation victim management and the haematologist in the future: time to revisit therapeutic guidelines? *Int. J. Radiat. Biol.*, 86: 636-648.

[3] Hérodin F., Drouet M. (2005) Cytokine-based treatment of accidentally irradiated victims and new approaches. *Exp. Hematol.*, 33: 1071-1080.

[4] Chao N. J. (2007) Accidental or intentional exposure to ionizing radiation: biodosimetry and treatment options. *Exp. Hematol.*, 35: 24-27.

[5] Ferrara J. L. M., Levy R., Chao N. J. (1999) Pathophysiologic mechanisms of acute graft-vs.-host disease. *Biol. Blood Marrow Transplant.*, 5: 347–356.

[6] Hayashi T., Morishita Y., Kubo Y., Kusunoki Y., Hayashi I. et al. (2005) Long-term effects of radiation dose on inflammatory markers in atomic bomb survivors. *Am. J. Med.*, 118: 83-86.

[7] *Calvi L. M., Adams G. B., Weibrecht K. W., Weber J. M., Olson D. P. et al. (.2003) Osteoblastic cells regulate the haematopoietic stem cell niche. Nature, 23: 841-846.*

[8] Friedenstein A. J., Chailakhyan R. K., Latsinik N. V., Panasyuk A. F., Keiliss-Borok I. V. (1974) Stromal cells responsible for transferring the microenvironment of the hemopoietic tissues. Cloning in vitro and retransplantation in vivo. *Transplantation*, 17: 331–340.

[9] Uccelli A., Moretta L., Pistoia V. (2008) Mesenchymal stem cells in health and disease. *Nat. Rev. Immunol.*, 8: 726-36.

[10] Koç O. N., Gerson S. L., Cooper B. W., Dyhouse S. M., Haynesworth S. E. et al. (2000) Rapid hematopoietic recovery after coinfusion of autologous-blood stem cells and culture-expanded marrow mesenchymal stem cells in advanced breast cancer patients receiving high-dose chemotherapy. *J. Clin. Oncol.*, 18: 307–316.

[11] Le Blanc K., Frassoni F., Ball L., Locatelli F., Roelofs H. et al. (2008) Developmental Committee of the European Group for Blood and Marrow Transplantation. Mesenchymal stem cells for treatment of steroid-resistant, severe, acute graft-versus-host disease: a phase II study. *Lancet*, 371: 1579–1586.

[12] Caplan A. I. (2009) Why are MSC therapeutic? New data: new insight. *J. Pathol.*, 217: 318-324.

[13] Pittenger M. F., Mackay A. M., Beck S. C., Jaiswal R. K., Douglas R. et al. (1999) Multilineage potential of adult human mesenchymal stem cells. *Science*, 284: 143–147.

[14] Phinney D. G., Prockop D. J. (2007) Concise review: mesenchymal stem/ multipotent stromal cells: the state of transdifferentiation and modes of tissue repair current views. *Stem Cells*, 25: 2896–2902.

[15] Nauta A. J., Fibbe W. E. (2007) Immunomodulatory properties of mesenchymal stromal cells. *Blood*, 110: 3499–3506.

[16] Dominici, M. et al. (2006) Minimal criteria for defining multipotent mesenchymal stromal cells. The International Society for Cellular Therapy position statement. *Cytotherapy*, 8: 315-317.

[17] Lange C., Schroeder J., Lioznov M. V., Zander A. R. (2005) High-potential human mesenchymal stem cells. *Stem Cells Dev.,* 14: 70-80.

[18] Lange C., Brunswig-Spickenheier B., Cappallo-Obermann H., Eggert K., Gehling U. M. et al. (2011) Radiation rescue: mesenchymal stromal cells protect from lethal irradiation. *PLoS One,* 5: 6(1):e14486.

[19] Graf T., Enver T. (2009) Forcing cells to change lineages. *Nature,* 462: 587-594.

[20] Dzierzak E., Speck N. A. (2008) Of lineage and legacy: the development of mammalian hematopoietic stem cells. *Nature Immunol.*, 9: 129–136.

[21] Krause D. S., Theise N. D., Collector M. I., Henegariu O., Hwang S. et al. (2001) Multi-organ, multilineage engraftment by a single bone marrow-derived stem cell. *Cell,* 105: 369–377.

[22] Dallas N. A., Samuel S., Xia L., Fan F., Gray M. J. et al. (2008) Endoglin (CD105): a marker of tumor vasculature and potential target for therapy. *Clin. Cancer Res.,* 14: 1931-1937.

[23] Fonsatti E., Maio M. (2004) Highlights on endoglin (CD105): from basic findings towards clinical applications in human cancer. *J. Transl. Med.,* 2: 18.

[24] Lange C., Kaltz C., Thalmeier K., Kolb H. J., Huss R. (1999) Hematopoietic reconstitution of syngeneic mice with a peripheral blood-derived, monoclonal CD34-, Sca-1+, Thy-1(low), c-kit+ stem cell line. *J. Hematother. Stem Cell Res.,* 8: 335–342.

[25] Huss R., Lange C., Weissinger E. M., Kolb H. J., Thalmeier K. (2000) Evidence of peripheral blood derived, plasticadherent CD34(-/low) hematopoietic stem cell clones with mesenchymal stem cell characteristics. *Stem Cells*, 18: 252–260.

[26] Rombouts W. J., Ploemacher R. E. (2003) Primary murine MSC show highly efficient homing to the bone marrow but lose homing ability following culture. *Leukemia,* 17: 160-70.

[27] Morikawa S., Mabuchi Y., Kubota Y., Nagai Y., Niibe K. et al. (2009) Prospective identification, isolation, and systemic transplantation of multipotent mesenchymal stem cells in murine bone marrow. *J. Exp. Med.,* 206: 2483-2496.

[28] Simmons P. J., Przepiorka D., Thomas E. D., Torok-Storb B. (1987) Host origin of marrow stromal cells following allogeneic bone marrow transplantation. *Nature,* 328: 429–432.

[29] Anderson D. J., Gage F. H., Weissman I. L. (2001) Can stem cells cross lineage boundaries? *Nature Med.,* 7: 393–395.

[30] Théry C., Ostrowski M., Segura E. (2009) Membrane vesicles as conveyors of immune responses. *Nat. Rev. Immunol.*, 9: 581-593.

[31] Remberger M., Sundberg B. (2004) Cytokine production during myeloablative and reduced intensity therapy before allogeneic stem cell transplantation. *Haematologica,* 89: 710-716.

[32] Lee R. H., Pulin A. A., Seo M. J., Kota D. J., Ylostalo J. et al. (2009) Intravenous hMSC improve myocardial infarction in mice because cells embolized in lung are activated to secrete the anti-inflammatory protein TSG-6. *Cell Stem Cell*, 5: 54-63.

[33] Lange C., Tögel F., Ittrich H., Clayton F., Nolte-Ernsting C., Zander A. R., Westenfelder C (2005) Administered mesenchymal stem cells are renoprotective in ischemia/reperfusion acute renal failures in rats. *Kidney Int.,* 68: 1613-1617.

[34] Tögel F., Weiss K., Yang Y., Hu Z., Zhang P. et al. (2007) Vasculotropic, paracrine actions of infused mesenchymal stem cells are important to the recovery from acute kidney injury. *Am. J. Physiol. Renal Physiol.*, 292: F1626-1635.

[35] Ittrich H., Lange C., Toegel F., Zander A. R., Dahnke H., Westenfelder C., Adam G., Nolte-Ernsting C. (2007) In-vivo magnetic resonance imaging of iron oxide labeled, arterially injected mesenchymal stem cells in kidneys of rats with acute ischemic kidney injury: Detection and monitoring at 3T. *J. Magn. Reson. Imaging*, 25:1179-1191.

[36] Tögel F., Hu Z., Weiss K., Isaac J., Lange C., Westenfelder C. (2005) Administered mesenchymal stem cells protect against ischemic acute renal failure through differentiation-independent mechanisms. *Am. J. Physiol. Renal Physiol.*, 289: F31-42.

[37] Meirelles Lda S., Fontes A. M., Covas D. T., Caplan A. I. (2009) Mechanisms involved in the therapeutic properties of mesenchymal stem cells. *Cytokine Growth Factor Rev.*, 20: 419-427.

[38] Caplan A. I., Dennis J. E. (2006) Mesenchymal stem cells as trophic mediators. *J. Cell Biochem.*, 98: 1076-1084.

[39] Lange C., Cakiroglu F., Spiess A. N., Cappallo-Obermann H., Dierlamm J. et al. (2007) Accelerated and safe expansion of human mesenchymal stromal cells in animal serum-free medium for transplantation and regenerative medicine. *J. Cell Physiol.*, 213: 18-26.

[40] Lange C., Brunswig-Spickenheier B., Eissing L., Scheja L. (2012) Platelet lysate suppresses the expression of lipocalin-type prostaglandin D2 synthase that positively controls adipogenic differentiation of human mesenchymal stromal cells. *Exp. Cell Res.*, 318: 2284-2296.

[41] Lange S., Zander A. R., Lange C. (2010) The MSC' in vitro community: where to go? Cellular Therapy and Transplantation (CTT) 1, No. 4; doi: 10.3205/ctt-2010- en-000051.01.

[42] Tögel F. E., Westenfelder C. (2010) Mesenchymal stem cells: a new therapeutic tool for AKI. *Nat. Rev. Nephrol.*, 6: 179-183.

In: Biological Responses, Monitoring and Protection … ISBN: 978-1-63463-852-4
Editor: K. P. Mishra

Chapter 12

HERBAL BASED STRATEGY FOR RADIOPROTECTION AGAINST RADIATION EXPOSURE

***Prabodha Kumar Meher*[1], *Prerna Sharma*[1] *and Kaushala Prasad Mishra*[1,2,*]**
[1]Division of Life Sciences, Research Center,
Nehru Gram Bharati University, Allahabad, India
[2]Ex-Head, Radiation Biology and Health Sciences Division,
Bhabha Atomic Research Center, Mumbai, India

ABSTRACT

Radiation exposure from events like accidents in hospitals, nuclear power plant catastrophe such as Chernobyl, Fukushima; nuclear weapon testing fallout from cold war era and use of atom bomb in war e.g. dropping of atom bomb on Hiroshima and Nagasaki in World War II poses threat to human health. In past decades, ionizing radiation has been increasingly used in research, medicine, agriculture and industry. Undesirable effects on health in many of these applications continue to be a matter of concern. Exposure of living cells to radiation generates reactive oxygen species (ROS) inside cells and thereby induces oxidative stress (OS) leading to various OS related diseases. Apart from ROS generation, radiation is known to cause DNA strand breaks, chromosomal aberrations, mutations and triggering apoptosis in the normal as well as in tumor cells. Extensive research has shown the ability of herbal drugs to protect normal cells from the ionizing radiation. Results from *in vitro* and *in vivo* studies have also shown that some herbal drugs have ability to act as anticancer drug in cancer radiotherapy but effectively protect normal cells against radiation damage. This review aims to discuss the mechanisms of action of herbal drugs in reduction of oxidative stress, repair of DNA damage , prevention of lipid peroxidation (LP) and activation of anti apoptotic processes by radiation in normal and tumor cells. A few examples are given concerning the mechanism of action of herbal formulations which point to potential of their practical applications in radioprotection and radiotherapy.

* Corresponding author email: mishra_kaushala@rediffmail.com, pkm046@gmail.com.

Keywords: Nuclear accidents, radiation exposure, health effects, radioprotectors, herbals

1. INTRODUCTION

Radiation existed on the earth from the day of its inception and living beings have evolved in the ocean of radiation environment [1]. However, it is remarkable to note that within a few tens of decades of discovery of radiation and radioactivity, a variety of applications have become common in diverse fields like medicine, industry, agriculture, electricity generation, archeological applications and in various consumer products [2]. In medical field, many diagnostic and therapeutic applications using radiation technology have found wide acceptance in public health service protocols. The focus of many current research programs is to develop newer applications in health domain but concern for safety and protection from harmful effects of radiation continues to be hot topic of investigation. The major efforts are devoted to prevent unwanted radiation exposure to the normal cells and organs. The major challenge is to spare the normal cells from radiation damage but effectively employ radioisotopes and radiation in care and improvement of human health [3,4]. It is noteworthy that currently about 17% of world wide electricity is being generated from nuclear power technology which contributes to improving the quality of life of people [2]. However, the events of Fukushima Daiichi Nuclear Power Plant disaster in March, 2011, Chernobyl nuclear plant catastrophe in April, 1986 and the three miles island accident on March, 1979 have legitimately raised new issues on environmental safety and public health concern from nuclear power plant. The Chernobyl accident set about 400 times more radioactive material into the environment compared to the atomic bomb explosion on Hiroshima in the year 1945 [5]. Numerous studies have reported various health consequences of nuclear disasters. The effects like mental sickness [6] thyroid cancer [7], leukemia and germ line mutations [8,9] are some of the major health consequences found among the populations living in and around radiation exposed areas. On the other hand, use of radionuclides in industry, agriculture and various consumer products is growing increasingly For example, in industries high penetrating γ rays are used for various applications viz. in metal casting, in measurement of level of fluid in beverage cans, density of tobacco in cigarettes etc. Applications of radioisotopes in agriculture as tracers in fertilizer research, sterile insect technique (SIT) to exterminate insects [2] and developing hybrid seeds are some notable examples.

It is however reasonable to anticipate the possibility of accidental over exposure to radiation which may be beyond human control. Therefore, it is highly essential to establish damage mitigating approaches in prior, during and post disaster situations to minimize the hazardous health effects from radiation exposure. In this context, efforts to shorten exposure time, controlling distance of exposure, immediate medical treatment, appropriate safety awareness and prevention of spread of contamination are some of the prime initiatives to be undertaken. Among these, medicinal treatment is one of the vital strategies to mitigate health effects of radiation exposure. The search for herbal drugs which can be safely given to victims to reduce the health effects of radiation and protect the normal cell on its intake before, during and after exposure are most urgent and highly warranted.

A wide range of synthetic and natural compounds have been studied to prevent the damage caused by radiation exposure [9,10]. However, large number of synthetic compounds had to be abandoned due to their acute toxicity, side effects and unaffordable costs [11-13]. Therefore, it is research challenge to find effective radioprotectors which should be non-toxic, devoid of side effects and affordable cost. Extensive recent researches have suggested compounds or extracts from herbs that hold the promise to be an efficient radioprotector. We have made an attempt to present a highlight of recent progress with regard to mode of action and mechanisms of various herbal extracts and compounds as potent radioprotector against radiation exposure.

2. Sources and Consequences of Radiation Exposure

Radiation exposure and its health effects depend on the type of incident radiation. Ionizing radiations cause severe health effects comparative to non-ionizing radiation because former carry enough energy to easily break the chemical bonds of the molecules in the intervening medium and stripping off electrons from the atoms creating ionization in the medium.

The circumstances like accidents in nuclear reactors, mishandling in disposal of radionuclear wastes from hospitals and industries, undetected radioactive sources in various products, nuclear power plant catastrophe, accidents in hospitals, industries and nuclear weapon may lead to ionizing radiation exposure along with spread of radionuclides within the environment. The consequence of exposure depends on the type of radioisotopes and the amount released into environment. For instance, the Three Mile Island accident released radioactive Iodine (^{131}I) and radioactive gases i.e. 43,000 curies of krypton to the environment [14]. Likewise in the Chernobyl nuclear accident more than one billion curie radioactivity was released into the environment [15]. Most recent Fukushima Daiichi Nuclear Power Plant disaster released in environment about 160 PBq of ^{131}I, 18 PBq of ^{134}Cs and 15 PBq of ^{137}Cs as reported by Nuclear and Industrial Safety Agency (NISA).

In medical practice especially in cancer radiotherapy normal cell often get exposed to lethal doses of radiation. It is an important challenge to spare the normal cells while efficiently killing the cancerous cells in cancer radiotherapy. In addition, the chances of radiation exposure from industrial, agricultural and consumer products are real and unavoidable.

3. Cellular Effects of Ionizing Radiation

The process of biological damage arises as a consequence of radiation absorption in the atoms of the cells, tissues and body. Absorption of ionizing radiation by atoms/molecules in the living cells provides sufficient energy for electrons to escape from atoms or to break the joining chemical bonds in structural molecules. These ionized atoms or molecules begin cascade of reactions leading to damage in vital biomolecules and structures inside the cell resulting in loss of function.

Radiation energy either directly damages biomolecules like lipids, proteins and DNA or indirectly by generation of free radical in water inside the living cells. Compared to direct interaction of radiation with DNA and other biomolecules, greater amount of radiation absorption occurs in water molecules. The reason is that as water or cytoplasm makes up most of the cell volume and, therefore, greater amount of deposition of radiation energy occurs in water generating many reactive free radicals such as electrons, hydroxyl radiacal and hydroperoxy radical and hydrogen atoms which are called primary water free radicals in cellular milieu. Primary free radicals further react with other compounds to generate secondary radicals and thus propagate a chain reaction of free radicals which are highly damaging to the cellular systems. Radiation generated primary and secondary free radicals are mainly the reactive oxygen species (ROS) and reactive nitrogen species (RNS). Cellular systems have several in-built defense machinery to neutralize the radiation generated ROS and RNS. However, when the equilibrium between the production of free radicals and their detoxification gets disturb, and generation free radical dominates inside the cell, it leads to the stressed condition within the cell know as oxidative stress [16,17]. Thus the free radicals causes damages to vital biomolecules by lipid peroxidation (LP), damage to DNA, protein, cell membrane [18,19], disruption of mitochondrial membrane potential, dysfunction of organelles [20], loss of signaling mechanisms [21] and triggers cell apoptosis. After irradiation, several mechanisms get activated in response to radiation action e. g. DNA repair mechanism, intracellular metabolic redox reactions, cell cycle checkpoints controls, intra and intercellular signaling cascades, apoptosis in damaged cells, mitotic linked cell death, and sometimes senescence too. However, if damage to DNA remained unrepaired, it may lead to the mutation in genetic material, loss in regulation of cell division followed by transformation of normal cells into tumor cell, uncontrolled cell growth and malignancy due to molecular, genetic and epigenetic processes in cells and tissues. However, biological damage processes may take many years to manifest into various health disorders.

4. Potential of Herbals in Radioprotection

4.1. Endogenous Defense by Antioxidants and Enzymes

Human body has internal defense systems to combat free radicals (ROS & RNS) generated by exposure to ionizing radiation. Each living cell has endogenous enzymatic and non-enzymatic defense mechanisms to neutralize free radicals and save them from harmful effects of oxidative stress. They include enzymes superoxide dismutase (SOD), catalase (CAT) and glutathione peroxidase (GPX) and others. If the endogenous defense system in the cell becomes inadequate to overcome the ROS generated by radiation, it becomes necessary to aid the defense system by providing external support through intake of some dietary antioxidants i.e. vitamins and micronutrients. These antioxidants neutralize free radicals at different stages of free radical mediated mechanisms like prevention, interception and repair. The cellular enzymes and antioxidants prevent the action of stress inducing mechanisms by converting the induced free radicals into inactive entities e.g. SOD catalyses dismutation of superoxide to hydrogen peroxide. The mechanism of interception includes scavenging of formed free radicals by the cellular enzymes and vitamins. A number enzymes, co-enzymes, isoenzymes of the cellular system take part in the repair of damage due to oxidative stress.

4.2. Prevention of Damage by Synthetic Drugs

Extensive research in the past decades has reported a large number of synthetic compounds which have ability to protect cells and tissues against damage by radiation exposure. Those containing sulphur have the most protective effects viz. aminothiols, aminodisulphides, tiourei derivatives, thiosulphur and thiophosphoric acid, dithiocarbamates, thiazole, some biogen amines and their derivatives [22-24]. Within the large number of available compounds, the most examined synthetic radiprotector were the cisteamine derivatives [25]. Most of the synthetic radioprotective compound act by different mechanism at molecular level, physiological-biochemical level and organic level inside the cell [26]. At molecular level, they work by neutralising free radicals, binding with biological targets and forming disulphide bonds; at physiological and biochemical level, they function as loosening unproteinated disulphide bonds and with biochemical stroke of the compound and at organic level, these compounds works by stimulating various organelle of the cells. However, most of the synthetic drugs are toxic, produce side effects and are excessively expensive.

5. Herbal Drugs Prevent Radiation Damage

Due to the acute toxicity, side effects and expensive, these synthetic radioprotectants have not found practical applications [27]. Thus search for an effective, cheaper and non-toxic compounds with efficient radioprotective capability is the focus of research in many laboratories the world over. It is found that herbals hold the promise due to their easy accessibility and acceptability, non-toxic even at higher concentration, reduced side effects and mostly of low cost. Therefore, our research group has extensively investigated the potency of several compounds of plant kingdom at cell, cell line and animal level of research. A few examples are given demonstrating the herbal compounds as potential radioprotectors which needs further studies at different level of trials for optimization and practical applications.

5.1. Triphala

Triphala, an ayurvedic formulation has been used in Indian medicinal system 'Ayurveda' as it has liver protective, blood cleansing [28], age preventing and immunity improving properties [29]. It is also known to possess antiviral, anti- fungal and anti allergic properties [30], and help to improve blood circulation, blood pressure and reduces level of cholesterol [31]. It also improves digestion and assimilation beside having therapeutic use in constipation [32]. Lately, Triphala has been suggested as effective anticancer drug as it increases the level of ROS in tumor cells, induces apoptosis in tumor cells while spare the normal cells [33-34] It was also shown that triphala provide protection against radiation induced sickness and mortality [35] by scavenging the free radicals like $OH^{\bullet}$, $O^{2-\bullet}$, $NO^{\bullet}$[36].Thus, ability to prevent normal cells because of its antioxidant property, considerable interest has developed to investigate the radioprotective ability of Triphala. Further, results have presented a significant findings that oral feeding of triphala pre and post irradiation, reduced mortality by 60% in case of whole body irradiation at 7.5 Gy. It is reported to prevent the DNA damage, significantly reduces activity of xanthine oxidase and increases superoxide dismutase activity which lowers the level of ROS in normal cells [34]. Therefore, Triphala may be a potent radioprotector for practical use and clinical studies are highly warranted.

5.2. Nigella Sativa

Nigella sativa (NS) commonly known as black cumin in Indian and Asian dietary additive has been shown as a potent herbal plant. Though it has been in used for various disease treatment like Asthma, Diarrhoea and Dyslipidemia for some time, recent studies from our laboratory on ethanolic extract of NS seed revealed remarkable radioprotection property of NS. Feeding 1000 mg/kg bw of ethanolic extract of NS to mice for 5days and 2hr after the last feeding when mice was whole body irradiated with 7.5 Gy γ radiation, reduction in radiation induced mortality and after 30 days 20% survival in the mice model was observed. Furthermore, strong antioxidant capacity, good oxidative damage inhibitor, increased blood leucocytes count, excellent potential in reducing apoptotic cell death and increased survival ability of the animal after radiation exposure was observed in the *in vivo* studies on mice fed with NS extract [37]. To mimic the clinical situation of cancer radiotherapy that needs simultaneously killing of cancer cells and protection to normal cells, studies are in progress. Study on potential of macerated extract of NS on transplanted fibro sarcoma showed differential effect on normal and tumor cells. [38] Radiation exposure escorts membrane oxidative damage via ROS and lipid peroxidation.

5.3. Eugenol

Eugenol, a natural antioxidant from Indian medicinal plants clove (*Syzygium aromaticum*) and tulsi (*Oscimum sanctum*) showed significant inhibition in malondialdehyde formation and increase in phospholipase C activity in irradiated liposomes [39]. Similarly, hydroalcoholic extract of *Aegle marmelos* fruit has shown protection against radiation-induced sickness and mortality in mice by decreasing the lipid peroxidation and increasing the enzyme glutathione [36]. In another study, it was found that aqueous extract of *Terminalia chebula* inhibited γ-radiation-induced lipid peroxidation in microsomes and caused damage to superoxide dismutase enzyme [40].

5.4. Silibinin and Thiol Antioxidant

Studies on silibinin, a non-toxic bioactive flavonoid compound found in plant milk thistle and consumed worldwide as a dietary antioxidant showed marked protection against radiation exposure. Generally, radiation exposure on cells induces DNA damage and micronucleus formation and silibilnin offered protection to lymphocyte DNA by reduction in DNA damage and micronuclei formation that mainly mediated by intracellular reactive oxygen species reduction. In addition, on feeding 70 mg/kg bw of silibinin for 3 days to mice and then irradiating mice with 7.5 Gy whole-body γ-exposure showed significant protection to DNA damage, blood leukocytes and radiation-induced mortality [28]. Similar studies on potential of different thiol antioxidants on radiation exposure have revealed that protection to DNA damage was concentration dependent. Among the thiol antioxidants *N*-acetyl cysteine, glutathione and thioproline; glutathione showed maximum protection and thioproline showed least protection to radiation induced DNA damage [41].

6. New Perspectives in Radioprotection Research

It is seen that major research efforts have been devoted to search for compounds which have ability to counter the reactive chemical species generated by ionizing radiation. However, the observed protection against damage does not fully explain the magnitude of prevention of damage by potential antioxidant compounds on the basis of their radical scavenging abilities. A viewpoint is growing concerning the mechanism of potential antioxidants their ability to accelerate or support repair of damage to DNA by radiation. Therefore, it looks rational to identify those antioxidants which have both radical scavenging and DNA repair capabilities. It would be prudent to focus future research in this direction which may help to discover efficient radioprotectors for practical situations.

It is further suggested that in view of our deeper understanding of the mechanisms of apoptosis in normal and cancer cells, it may be possible to find specific drugs which may prevent apoptosis in normal cells after radiation exposure but induce apoptosis in tumor cells following irradiation. Therefore, greater effort is needed to search for compounds which may play pro-survival role in normal cells perhaps by inhibiting the apoptotic pathways in response to ionizing radiation. We believe that search for herbal compounds from plant sources with anti-apoptotic properties in normal stressed cells may turn out to be a promising endeavor in future.

Summary and Future Research

With the diverse uses of radiation, the possibilities of unwanted radiation exposure or accident cannot be avoided. In this scenario, mitigating the effects of radiation by use of radioprotectors appears to be a promising approach for practical applications. In order to get desired protection the cascade of mechanisms induced by radiation need to be identified and controlled. Studied have shown various synthetic and herbals drugs as potential radioprotectors. However, recent studies reports compounds or extracts from herbals are more promising due to their nontoxic, moderate side effect and cost effectiveness. Many of them works by reducing the intra cellular ROS formation, inhibition of lipid peroxidation, DNA damage, micronucleus formation and ultimately inhibiting cellular apoptosis. With current understanding, a drug that can act as apoptosis inhibitor may turn out an ideal radiprotector.

However, only few herbs have been explored and results from clinical trials are lacking. Research has however set out the direction for search of herbals which may have pro-survival and anti-apoptotic properties. It is evident that further extensive investigations and clinical trials are warranted.

ACKNOWLEDGMENTS

Prabodha Kumar Meher and Prerna Sharma acknowledge the financial support as Senior Research Fellow from Board of Research in Nuclear Sciences (BRNS), Dept. of Atomic Energy, Govt. of India.

REFERENCES

[1] Hall E.J. "Radiation and Life" http://www.world-nuclear.org/education/ral.html accessed on 24-12-2012.

[2] Uses of radiation, http://www.hko.gov.hk/education/dbcp/radiation/eng/r15.htm accessed on 27-03-2014.

[3] Durante M. & Loeffler J. S. (2010) Charged particles in radiation oncology, *Nat. Rev. Clin. Oncol.* 7, 37–43.

[4] Girdhani S, Bhosle S. M., Thulsidas S. A., Kumar A, Mishra K. P. (2005). Potential of radiosensitizing agents in cancer chemo-radiotherapy. *J. Can Res. Ther,; Journal of cancer research and therapeutics*, *1*(3), 129-31.

[5] International Atomic Energy Agency (IAEA), Facts: The accident was by far the most devastating in the history of nuclear power; Ten years after Chernobyl: What do we really know? http://www.iaea.org/Publications/Booklets/Chernoten/facts.html accessed on 01 April 2014.

[6] Bromet EJ, Havenaar JM, Guey LT (2011),A 25 year retrospective review of the psychological consequences of the chernobyl accident, *Clin. Oncol.* (R Coll Radiol). 2011 May;23(4):297-305.

[7] Tronko M.D., Bogdanova T.I., Komissarenko I. V., Epstein O.V., Oliynyk V., Kovalenko A., Likhtarev I.A., et. al., (1999), Thyroid carcinoma in children and adolescents in ukraine after the Chernobyl nuclear accident, *Cancer* 86:1, 149–156.

[8] Dubrova Y.E., Nesterov V. N., Krouchinsky N.G., Ostapenko V.A., Neumann R., Neil D.L. & Jeffreys A.J., (1996), Human minisatellite mutation rate after the Chernobyl accident, *Nature,* 380, 683-686 doi:10.1038/380683a0.

[9] Dubrova Y.E., Nesterov V. N., Krouchinsky N.G., Ostapenko V.A., Vergnaud G., Giraudeau F., Buard J., Jeffreys A.J., (1997) Further evidence for elevated human minisatellite mutation rate in Belarus eight years after the Chernobyl accident, *Mutation Research* 381, 267–278.

[10] Pandey, B.N., Mishra, K.P., 2004. Modification of thymocytes membrane radiooxidative damage and apoptosis by eugenol. *J. Environ. Pathol. Toxicol. Oncol.* 23, 117–122.

[11] Weiss, J.F., Landauer, M.R., 2000. Radioprotection by antioxidants. *Ann. NY Acad. Sci.* 899, 44–60.

[12] Maurya D.K., Paul T., Devasagayam A. (2011). Role of Radioprotectors in the Inhibition of DNA Damage and Modulation of DNA Repair After Exposure to Gamma-Radiation, Selected Topics in DNA Repair, Prof. Clark Chen (Ed.), ISBN: 978-953-307-606-5, InTech.

[13] Landauer M.R., Davis H.D., Dominitz J.A., Weiss J.F. (1987). Dose and time relationships of the radioprotector WR-2721 on locomotor activity in mice. *Pharmacol. Biochem. Behav.* 27: 573–576.

[14] Landauer M.R., McChesney D.G., Ledney G.D. (1997). Synthetic trehalose dicorynomycolate (S-TDCM): behavioural effects and radioprotection. *J. Radiat. Res.* 38:45–54.

[15] http://www.nrc.gov/reading-rm/doc-collections/fact-sheets/3mile-isle.html accessed on 02/04/2014.

[16] The Truth About Chernobyl, Critical Mass: Voices for a Nuclear-Free Future, Ruggiero and Sahulka, Eds., 1996, by Open Media, p. 127.

[17] Pandey B.N., Mishra K.P.,(2003), Oxidative membrane damage and its involvement in radiation induced apoptotic cell death, *Iran J. Radiat. Res.,* 1: 17-22.

[18] Pandey B.N., Mishra K.P.,(1999), Radiation induced oxidative damage modification by cholesterol in liposomal membrane. *Radiat Phys. Chem.*; 54:481-489.

[19] Fredickson J.K., Lis M., Gaidamakova E.K., Matrosova V.Y., Zhai M., Sulloneay H.M., Schlten J.C., Brown M.G., Balkwill D.L., Daly M.J.,(2008) Protein Oxidation: Key to Bacterial Desiccation resistance? *The ISME Journal* 2, 393–403; doi:10.1038/ismej.2007.116.

[20] Haimovitz- friedman A., Kan C.C., Ehlliter D., Perand R.S., McLovghlin M., Fuks Z., Kolesnick R.N., (1994) Ionizing radiation act on cellular membranes to generate ceramide and initiate apoptosis. *J. Exp.Med.* 180, 525-535.

[21] Spitz D.R., Azzam E.I., Li J.J., Guis D., (2004) Metabolic oxidation / reduction reactions and cellular responses to ionizing radiations: a unifying concept in stress response biology, *Cancer Metastasis, Rev*. 23,, 311-322.

[22] Giambarresi L. and Jacobs J.A., (1987) Radioprotectans. In: J.J.Conklin and I.R.Walker (eds) Military Radiobiology, Academic Press, Orlando, 265-301.

[23] Mishra S., Malhotra P., Gupta A.K., Singh P.K., Mishra A.K., Javed S., Kumar R. (2014) Novel method for screening of radioprotective agents providing protection to DNA ligase against gamma radiation induced damage, *International Journal of Radiation Biology,* 90:2, 187-192 (doi:10.3109/09553002.2014.868613).

[24] Devi P.U., Bisht K.S., Vinitha M. (1998), A comparative study of Ocimum flavonoids and synthetic aminothiol protectors in the mouse, *The British Journal of Radiology*, 71: 782-784.

[25] Milosevic Z., Kljajic R., and Herceg Z., (1990)Efficiency of the phosphortioate compond, WR-2721, as a radioprotective agent in lethally irradiated pigs. YRPA, Proceeding of the 3-rd Italian-Yugoslav Symposium on Radiation Protection, 36-43, Plitvice, Yugoslavia.

[26] Yuhas J.M. and Philips T.L., (1982) Pharmacokinetics and mechanisms of action of WR-2721 and other protective agents. Radioprotector and Anticarcinogenesis. New York: Academic Press, 639-653.

[27] Landauer M.R., Davis H.D., Kumar K.S., Weiss J.F., (1992) Behavioral toxicity of selectedradioprotectors, *Adv. Space Res.* 12: 273–283.

[28] Tiwari P., Kumar A., Ali M., Mishra K.P., (2010) Radioprotection of plasmid and cellular DNA and Swiss mice by silibinin, *Mutation Research/Genetic Toxicology and Environmental Mutagenesis* 695: 55–60.

[29] Juss S.S., Triphala-the wonder drug. *Indian Med. Gaz.*,1997, 131: 194-96.

[30] Singh P.K., Mycotoxin elaboration in triphala and its constituents, *Indian Phytopathol*, 2003, 56, 380-383.

[31] Thakur C.P., Thakur B., Singh S., Sinha P.K., Sinha S.K., (1998) The Ayurvedic medicines haritaki, amla and bahira reduce cholesterol induced Atherosclerosis in rabbits, *Int. J. of Cordiol*, 21: 167-175.

[32] Nadkarni A.K., Indian Meteria Media, Popular Press Ltd., Mumbai,1976,3rd edu. Pp.1308-1315.

[33] Sandhya T, Lathika K.M., Pandey B.N., Mishra K.P., (2005), Potential of traditional ayurvedic formulation, Triphala, as a novel anticancer drug, *Cancer Lett.* 231: 206–214.

[34] Sandhya T., Mishra K.P., Cytotoxic response of breast cancer cell lines, MCF 7 and T 47 D to Triphala and its modification by antioxidants (2006), *Cancer Lett.* 238: 304–313.

[35] Jagetia G.C., Baligaa M.S., Malagib K.J., Sethukumar K. M., The evaluation of the radioprotective effect of Triphala (an ayurvedic rejuvenating drug) in the mice exposed to γ-radiation, (2002) *Phytomedicine,* Volume 9:2, Pages 99–108.

[36] Jagetia G. C., Venkatesh,P., & Baliga M. S. (2004). Fruit extract of *Aegle marmelos* protects mice against radiation-induced lethality. *Integrative Cancer Therapies*, 3(4), 323-332.

[37] Rastogi L., Feroz S., Pandey B.N., Jagtap A., Mishra K.P., (2010), Protection against radiation induced oxidative damage by an ethanolic extract of *Nagilla sativa* L., *Int. J. radiat. Biol.* 86:9, 719-731.

[38] Velho-Pereira R., Kumar A., Pandey B.N., Mishra K.P. and Jagtap A. G., (2012), Radioprotection by Macerated Extract of *Nigella sativa* in Normal Tissues of Fibrosarcoma Bearing Mice, *Indian J Pharm Sci.* 74(5): 403–414. doi: 10.4103/0250-474X.108415.

[39] Pandey B.N., Lathika K.M., Mishra K.P., (2006), Modification of radiation-induced oxidative damage in liposomal and microsomal membrane by eugenol, *Radiation Physics and Chemistry* 75: 384–391.

[40] Naik G.H, Priyadarsini K.I., Naik D.B., Gangabhagirathi R., Mohan H., (2004) Studies on the aqueous extract of Terminalia chebula as a potent antioxidant and a probable radioprotector, *Phytomedicine*, 11: 530–538.

[41] Tiwari P., Kumar A., Balakrishnan S., Kushwahaa H.S., Mishra K.P. (2009), Radiation-induced micronucleus formation and DNA damage in human lymphocytes and their prevention by antioxidant thiols, *Mutation Research* 676:62–68.

In: Biological Responses, Monitoring and Protection … ISBN: 978-1-63463-852-4
Editor: K. P. Mishra

Chapter 13

Role of Antioxidants in Cellular Response to Ionizing Radiation

Renu Dayal, Amrita Singh and K. P. Mishra*

Division of Life Sciences, Research Centre,
Nehru Gram Bharati University, Allahabad, India

Abstract

Exposure of cells to ionizing radiation IR) induces generation of reactive oxygen species (ROS) which may damage vital cellular molecules if not detoxified. Many herbs are known to protect cells from deleterious effects of induced ROS due to their antioxidant (AOs) properties. It is observed that endogenous antioxidants like SOD, GSH and GPX have potential to neutralize the stress produced due to ROS resulting in promotion of cellular survival. This paper describes the prospects of searching potent herbal antioxidants which possess radioprotective capability and their influence on initial and long term radiation damage to cells after radiation exposure. Furthermore, they can be regularly consumed as dietary component. Although the exact mechanism is unclear but it seems possible that these AOs recognize the cytosolic redox status of cells which is known to be different in normal and stressed cells. Another possibility is that AOs up regulate endogenous defense machinery reflected in enhanced intracellular SOD or GSH. Many AOs from herbal sources have been reported to produce protective effect in tissues exposed to radiation. The molecular mechanism is reported to involve their ability to down regulate ROS below toxic levels and consequent protection against radiation.

Keywords: Radiation, reactive oxygen species, herbals, antioxidants, radioprotection

1. Introduction

Living beings co-exists in an environment that has significant levels of ionizing and non-ionizing radiations. Humans of present industrial era are exposed to one or other forms of this

* Corresponding author: mishra_kaushala@rediffmail.com, renu.dayal10@gmail.com.

ubiquitously existing ionizing radiation. Although radiation is a boon, satisfying the needs of energy, agriculture, modern medicine but on the other side, direct exposure to high doses of radiation poses threat to life. As a consequence of chain reaction initiated by radiation exposure, certain kinds of oxidants such as reactive oxygen species (ROS) are produced. In consequence, intracellular oxidative stress is generated depending on the dose and duration of the exposure. In a typical cellular response to which cell leaves the course of cell cycle and enters into G0 phase. Moreover, high doses of such ionizing radiation may cause damage to vital cellular organelles if not detoxified, leading to ill health effects.

Radiation damage to normal tissues might appear instantly or may take long time to appear. The type and intensity of damage depends on the cells and organs being exposed to radiation, the doses and the time during and after radiation exposure that is being assayed for adverse health effect (Averbeck,2010). Besides this, such prolonged and/or high dose radiation exposure, results in many other unpleasant health consequences like radiation sickness, cardiovascular diseases, skin diseases (Liu et al., 2008; Mettler, 2012;Giardi et al.,2013).While most of the damages expected to arise after irradiation can be ameliorated by conventional radioprotectors but they fail to do so in long terms. Antioxidants of herbal origin have arisen as promising candidate to strengthen the inbuilt defense system of the host to fight ill effects of radiation in long term. This review will outline some important radiation related toxicological processes and discuss the promising role of some dietary antioxidants in affecting cellular processes in terms of the health effects likely to arise from radiation exposure. The role that different herbals and their antioxidant component might play in preventing ionizing radiation induced damage will also be explored.

2. Radiation Oxidative Stress

The constant exposure to IR is mainly derived from outer space sources, medical field and nuclear field. Henceforth, people working or residing near such places are expected to be more prone to long term health consequences of low doses of IR. Exposure of eukaryotic cells to ionizing radiation (IR) results in the immediate formation of free radicals known as reactive oxygen species (ROS) such as superoxide ($^{\cdot}O^{-}_2$), hydroxyl radical ($^{\cdot}OH^{-}$), hydrogen peroxide (H_2O_2). The major sites for ROS generation are complexes I and III of mitochondria. The persistent production of ROS is capable enough to disturb the delicate redox balance of cell regeneration and death by depleting the level of endogenous AOs such as glutathione GSH), catalase (CAT). As a consequence of which oxidative stress arises, a critical condition that if not checked timely, may trigger subsequent alterations in multiple intracellular processes. Further, if ROS enters into the cytosol via oxidative burst, may interact with macromolecules such as DNA, protein, membrane lipids, leading to cell dysfunction and mortality through a series of molecular events. These processes usually comprised of redox sensitive signaling pathways, transcription factor activation, gene expression, and metabolic activities (Spitz et al., 2004). Oxidative damage and cellular dysfunction by absorption of IR over a prolonged period of time induces several diseases including cancer, cardiovascular and retinal diseases (Giardi et al., 2013).

2.1. Induced ROS Cause Oxidative Stress in Cells

The physiological manifestations of these radiation-induced alterations in redox sensitive processes have been suggested to contribute to adaptive responses, bystander effects, cell cycle perturbations, cytotoxicity, heat-induced radiosensitization, genomic instability, inflammation, and fibrosis (Spitz et al., 2004, Pandey et al., 2011). Figure 1 presents a systematic illustration of IR mediated oxidative stress and its harmful consequences. As discussed earlier, disturbed redox balance thus consecutive oxidative damage of any cellular constituent, significantly contributes to signaling pathway, disturbed enzymatic activities that causes mutation and subsequent diseases (Ruch et al., 1983; Manda et al., 2009; Hamanaka et al., 2010).

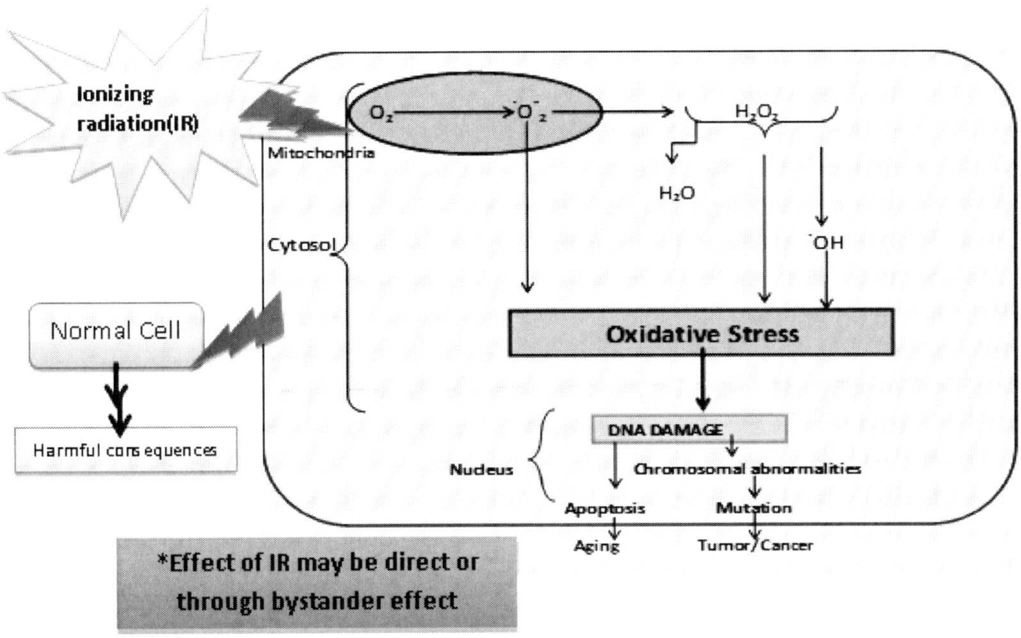

Figure 1. Schematic diagram showing the harmful effects of ROS induced oxidative stress in a mammalian cell (Symbols: O_2=Oxygen, $O^{\cdot}_2$=Superoxide radical; H_2O_2= Hydrogen peroxide; OH= Hydroxyl radical; H_2O=Water).

3. Mechanism of Radiation Damage to Cells

IR mediated ROS since highly reactive and unstable in nature, produces change in the molecule through which they pass, making them unstable. The ROS activated molecules further triggers expression of transcription factors like NF-κB, ERK and other downstream signaling factors(Riley, 1994; Kamata et al., 1999; Narang et al., 2009); likewise this chain of reaction continues. IR mediated damage may be targeted (directly on cell) or non-targeted comprised of indirect cellular events that is release of cytokines and bystander effect. Depending on the dose and duration of exposure radiation, stress response might be induced in two ways– (i) Short term, high dose radiation exposure- involve ROS mediated immediate stress response by signaling, that causes local radiation injuries, most signs and symptoms of which are similar to a thermal burn, (ii) Low dose, long term health issues- are non-targeted delayed radiation impacts through release of cytokines (bystander effect), the symptoms of those take time to appear (Averbeck, 2010; Pandey et al., 2011; Mettler, 2012). Cellular mechanisms governing redox homeostasis likely involve their integration with other stresses. Many genes induced in response to specific oxidant stresses have been identified and include transcription factors, replication proteins, proteases, protease inhibitors, proteins affecting cell proliferation and various antioxidants, i.e. hemeoxygenase, MT, and MnSOD(Janssen et al., 1993).

4. Cellular Targets for Radiation Action

ROS imposes threat to various macro- and micro molecules including DNA, cell membrane, lipids, proteins and mitochondria if not detoxified (Halliwell et al., 1989; Valko et al.,2007; Güngör et al., 2010). Mitochondria are the primary site of free radical generation; hence, they are the prime targets of radiation induced damage along with cell membrane. The mitochondrial ROS released into the cytosol contributes in significant damage to macromolecules. However, parameters such as DNA repair system, cell cycle checkpoints and some enzymes try to maintain the genomic integrity of stressed cells (Belli et al., 2002). Figure 2, provides a systematic overview of various targets of IR and expected effects.

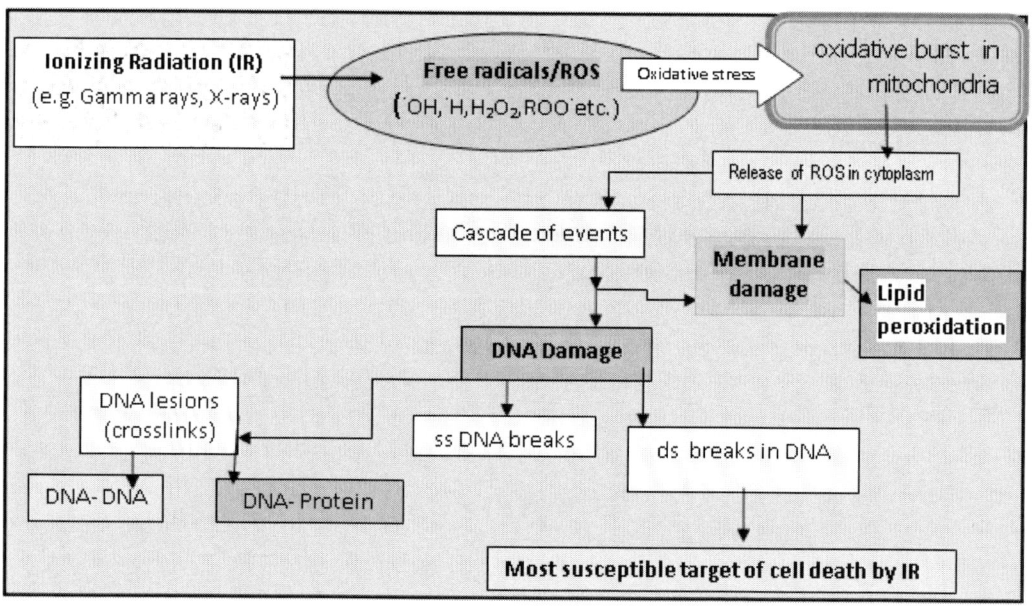

Figure 2. Schematic diagram showing various targets of ionizing radiation induced damage.

4.1. Mitochondria

Mitochondria are the primary site of ROS generation hence most vulnerable to oxidative stress and change in mitochondrial membrane potential. ROS are formed within mitochondria as a byproduct of respiratory chain (Manda et al., 2009). It is this leakage of electrons from the electron transport chain that promotes ROS mediated oxidative burst and further damage to mitochondria. Persistent production of ROS in irradiated cells may result from, depleted level of endogenous AOs, damage to components of the mitochondrial respiratory processes and an effect on plasma membrane oxidases (Burdon, 1996; Kamata, 1999). Up-regulation of ROS encourages early events of apoptotic cascade by inducing depolarization, loss in membrane potential and release of cytochrome c (Kim et al., 2004; Ott et al., 2007).

4.2. Oxidative Damage to Membranes

Cell membrane protein and lipids are also significant targets of free radical attack. Any change in these molecules increases the risk of cell death through apoptosis or necrosis. Henceforth, we propose such proteins and lipids to be the RADIATION CENTREs specifically present in radiation sensitive and transformed cells. Figure 3 shows a systematic flowchart depicting various modes of ROS induced lipid and protein damage.

4.2.1. Lipids

Membrane lipids of cellular organelles are highly susceptible to damage tempted by free radicals. ROS attacks the lipids (fatty acids) of cell membrane, as a consequence of which they are converted to polyunsaturated fatty acids (PUFA) e.g. lipid hydroperoxy radicals, lipid hypoperoxides (Güngör et al., 2010; Cai et al., 2012; Alvarez et al., 2012). The release of products formed during lipid peroxidation subsequently reacts with and alter cellular components, that may result in damage or release of membrane proteins (Okunieff, 2008).

4.2.2. Protein Damage

Oxidative modification of proteins can be of three types: modification of specific amino acid, free radical mediated peptide cleavage, and formation of protein cross-linkage due to reaction with lipid peroxidation products (Lobo et al., 2010). Oxidative stress causes cross linkage in the amino acid sequences of protein and fragmentation of polypeptide bonds (Stadtman, 2014). Such damage to protein products may affect the activity of enzymes, receptors, signal transduction mechanism, enzyme activity, heat stability, membrane transport and proteolysis susceptibility which leads to aging (Lobo et al., 2010).

4.3. DNA Damage and Repair

Damage to DNA causes modification of the purine and pyrimidine bases, the deoxyribose backbone, ruptured single and double strand breaks, as well as cross-links to other molecules (Valko et al., 2007; Ott et al., 2007) such as 8-Hydoxyguanodine (8-OHDG), a DNA lesion is formed from deoxyguanosine due to oxidation of its 8th carbon by free radicals. Double-strand breaks (DSB) are biologically important as they are most susceptible to radiation-induced damage (Belli et al., 2002). DNA damage can result in genomic instability, replication error, induction of signal transduction pathway. Acceleration of repair process will protect cells against radiation. The damage activates signaling pathways that ensures cell survival by activating checkpoints responsible for delay in cell cycle progression that allows more time for repair. Cells those at G1/S and S phases prevent replication of damaged DNA and those at G2/M phase prevent segregation of changed chromosomes (Belli et al., 2002). It was suggested that arrest at the S/G2M phases of the cell cycle may have had a radioprotection action (Khafif et al. 2005; Shimizu et al., 2005). Damage which is not repaired with high fidelity leads to chromosomal aberration, mitotic or necrotic death of the cell (Collis et al., 2004).

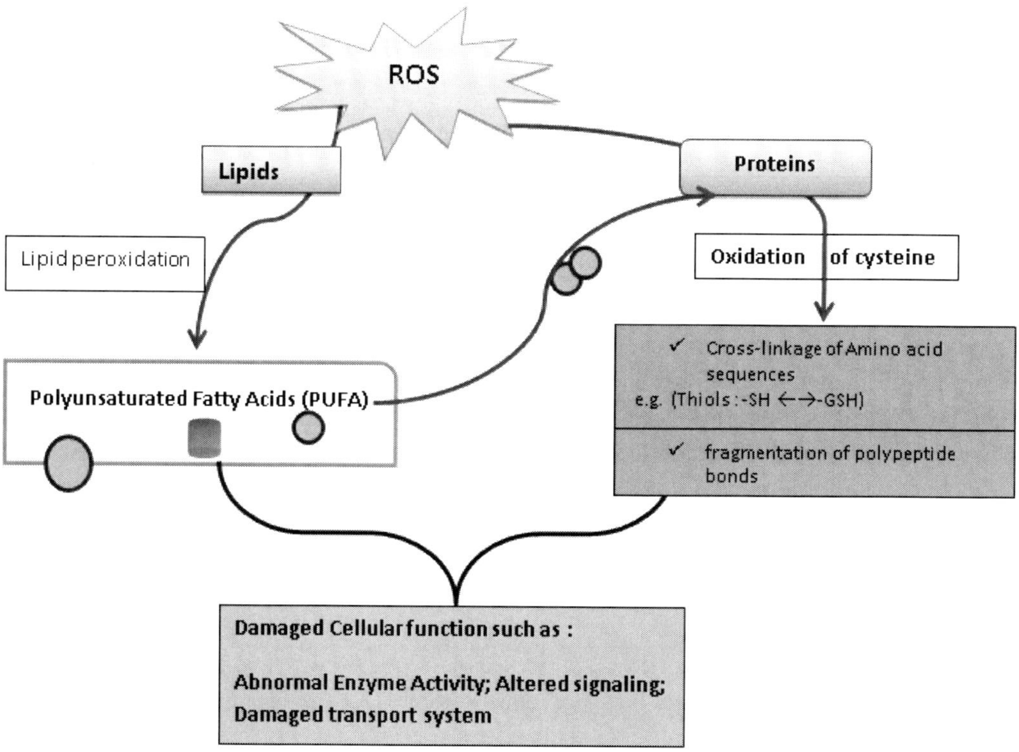

Figure 3. Schematic diagram depicting various modes of ROS induced lipid and protein damage.

5. Antioxidants Modify Cellular Damage

Antioxidants have unique adaptive systems that enable adaptation to oxidative environment/oxidative stress. Henceforth, understanding the pathways leading distinct antioxidant response to cellular environment will help in developing strategies to protect against radiation exposure. In their investigation on different in vivo models Limón-Pacheco et al. (2009) suggested, that organs and tissues contain distinct antioxidant systems, that may form the basis for differential susceptibility to environmental toxic agents (Limón-Pacheco et al., 2009). Bravard and coworkers demonstrated adaptive response of human lymphoblastic cells towards ionizing radiation by activation of antioxidant defense system (Bravard et al., 1999). However, results by MacDonald et al. (2010) indicated that the transcription factor NF-E2-related factor 2 (Nrf2) binds the antioxidant DNA response element (ARE) to activate important cytoprotective defense systems of cells. They demonstrated Nrf2-ARE pathway as an important factor that maintains resistance to irradiation, nevertheless it operates as a second-tier antioxidant adaptive response system activated by radiation only under specific circumstances (MacDonald et al., 2010). Also, in their study Eken and co-workers (2012) showed the presence of stimulant effect of chronic low-dose radiation in hospital staff occupationally exposed to ionizing radiation, resulting in enhanced resistance to oxidative stress, contributing to adaptive response towards ionizing radiation (Eken et al., 2012).

However, the fact is that, till now there isn't any effective radioprotector that possess null or least toxicity to normal cells at optimum protective doses. Most of them are not much effective at optimum protective dose levels, hence practically require high dose or are effective only when administered prior to irradiation(Pandey et al., 2010). It is therefore, important to develop strategies to improve radioprotection approaches to combat harmful effects of radiation with new therapeutic agents of herbal origin. Cumulative research findings suggest that indigenous medicinal plants and their bioactive compounds have immense potential to detoxify deleterious effects of ROS because of potent AO system within them. Some of the plants or their herbal compositions were found to be effective in inhibiting MMPS, NF-κB translocation, lipid peroxidation. It was found that pretreatment of various extracts prior to irradiation reduced the tumor volume, endogenous antioxidant level (GSH, MMPS, CAT etc.) of tumor cells. On the other hand, they accelerated the recovery of endogenous antioxidant level of normal tissues to average redox levels along with strong free radical scavenging activity (Denissova, 2012). From these findings we can portray the probability that AOs up-regulate endogenous defense machinery of normal cell towards irradiation induced stress like enhanced intracellular SOD or GSH.

6. Potent Antioxidants

Oxidative damage to cell components, DNA, proteins and lipids grounds for chronic diseases such as cancers, cardiovascular diseases, osteoporosis. Antioxidants present in food can help limit this damage by acting directly on reactive oxygen species or by stimulating endogenous defense systems. The phenolic groups in polyphenols can accept an electron to form relatively stable phenoxyl radicals, thereby disrupting chain oxidation reactions in cellular components. The antioxidant potency of polyphenols has been evaluated *in vitro* by measuring their ability to trap free radicals and reduce other chemicals.

Depending upon the nature of availability, antioxidants can be classified as synthetic and natural antioxidants, dietary in particular. The sulfur containing thiols are lead representatives of synthetic/conventional AOs, whereas natural antioxidants comprises of phytoconstituents or bioactive compounds purely organic in nature.

6.1. Thiols (-SH Compounds)

Early research identified sulfur-containing (-SH) antioxidants as radioprotectors with the most beneficial therapeutic ratio during irradiation. The relevance of thiols in protection has also been demonstrated by the fact that their high intracellular levels in mitotic cells leads them to become radioresistant (Capizzi and Oster, 2000; Sinclair, 1968). The first evidence of radioprotector is in form of cysteine derivatives by Patt and coworkers, that projected a new field of research in radiotherapy (Patt et al., 1949). Since, then a large number of sulphur/thiol containing compounds and their analogues such as aminothiol, amifostine (WR-1065), captopril, mesna have been tested for their validity as effective radioprotector. Even though these compounds have substantial toxicity when examined *in vivo* (Kouvaris et al., 2007; Grdina, 2009; Kemp et al., 1996). However, amifostine (WR-2721) with its active form WR-1065, is the first selective-target and broad-spectrum radioprotector clinically approved radioprotector used for the protection of normal cells neighboring the irradiated cancer cells (Kouvaris et al., 2007; Wong et al., 2010). Kemp and coworkers demonstrated that it induces antioxidant enzymatic activities in normal tissues and a transplantable tumor that can affect radiation response (Kemp et al., 1996). But that too failed in post irradiation protection, thus have many side effects and limited medical/clinical use. Nausea, vomiting, hypotension, and allergic reactions were the most common side effects (Kemp et al., 1996; Zabbarova and Kanai, 2008; Brizel, 2000) along with high cost, short time impact, failure to provide protection to normal cells and post-therapy effects (Brizel et al., 2000; Rades et al., 2004).

6.2. Dietary AOs

Toxic side effects of so far developed conventional (synthetic) radioprotectors limit their use in medical field. Conversely, antioxidants with herbals origin possess bioactive metabolites (AOs) that have potential to act differentially under different cellular environments and molecular mechanisms. Dietary/herbal antioxidants might open a new window of protection against lethal doses of radiation. For instance, concurrent use of plant derived alkaloid etoposide appears to be one of the most efficient concurrent treatment in combination with radiotherapy (Chard et al., 1978; Slater et al., 1986). It is supposed to increase radiosensitivity of cancer cells that divide more quickly than normal cells; cancer cells are more likely to be affected by etoposide than normal cells. Yet, they fail to discriminate cytosolic status between tumor and normal cells, hence unable to protect rapidly dividing normal cells neighboring the radiation targeted area.

In the past few years, use of plant and plant products (herbals) has gained admirable status as natural medicinal resource in almost all fields of medicine due to its effectiveness and reliability. In the present chapter, some major findings by our research group and other parallel investigations on promising radioprotective efficacy of indigenous herbals have been briefly summarized in a tabular form (Table 1). Results from our animal experiments indicated that antioxidant phytoconstituents, such as eugenol, ellagic acid and gallic acid (from Triphala), are protective against radiation lethality (Sandhya, 2006; Pandey et al., 2006; Bhosle et al., 2005). Significant results from many other laboratories have also proved radioprotective efficacy of natural antioxidants. Most herbal antioxidants and phytochemicals have the advantage of low toxicity, they are generally protective. Following are some potent dietary AOs displaying radioprotective potential, studied by various researchers inclusive of our research groups.

The alcoholic extract of seeds of *Nigella sativa* (black cumin) exhibited radioprotective effects through increased level of intracellular antioxidants like glutathione [GSH] (Rastogi et al., 2010; Velho-Pereria et al., 2012). Furthermore, significant free radical scavenging and protection against DNA damage in cell free systems was found.Moreover, our previous research findings on Triphala, an ayurvedic herbal formulation showed that the aqueous extract significantly inhibited oxidative damage hence protected normal cells against ionizing radiation. (Sandhya, Lathika et al., 2006).

Table 1. List of some herbal AOs with prospective radioprotection potential and their dietary sources

S.No.	Herbal Antioxidants	Dietary Source	Target	Mechanism	References
1.	Caffeic acid	Coffee propolis of honeybee hives	NF-κB MDA, XO and ADA	Inhibits transcription factor; Down-regulated MDA, XO and ADA with significant increase of endogenous AO	Akyol et al., 2012 Mansour et al., 2012
2.	Carotenoids	Tomato, Carrots, Spinach, Citrus fruits	Inflammatory genes	Modulate the expression of inflammation related genes in retinal pigment epithelial cells.	Narici et al., 2009
3.	Catechin Phenol	Black berries Black grapes Barley, Green tea	Caspases 3	Inhibit irradiation induced apoptosis	Kondo et al., 2004
4.	Curcumin	Turmeric	NF-κB, COX-2 protein	inhibit NF-kB activation inhibit COX- 2 expression by suppressing p38 MAPK and JNK activities	Zong et al., 2012 Cho et al., 2005
5.	Ellagic acid	Black berries, pecans, pomegranates, strawberries & walnuts	DNA	Reduce endogenous oxidative DNA damage by DNA repair mechanism	Aiyer et al., 2008
6.	Eugenol	Clove oil, nutmeg, cinnamon, basil & bay leaf	Cell membrane	Inhibit radiation induced membrane oxidative damage	Pandey and Mishra, 2004 Pandey et al., 2006
7.	Gallic acjd	blueberries, apples, tea leaves, oak bark, walnuts, TPL	Cell membrane; AKT; miRNA; NF-κB,	inhibited phosphorylation of AKT and down-regulation of miRNA, NFκB, increase apoptosis	Sandhya et al., 2006 Shi et al., 2008
8.	Gingerol	Ginger	COX-2 gene	inhibited induction COX-2 mRNA and protein, as well as NF-κB translocation Activation of caspase-3, -8, -9	Kim et al., 2007
9.	Organosulphur compounds (including –SH groups)	Garlic, Onions	Chromosomes	prevent chromosomal damage of normal tissues adjacent to the tumor	Singh et al., 1996
10.	Resveratrol	Blue berries Dark chocolate Red grapes	DNA	Protect DNA damage through radiation in normal cell	Denissova et al., 2012

Abbreviations: ADA, adenosine deaminase; AO, Antioxidants; CAT, catalase; COX-2, Cyclooxygenase-2; DNA, Deoxyribonucleic acid; ERK, extracellular signal regulated kinases; JNK, c-Jun N-terminal kinases; MAPK, mitogen activated protein kinases; MDA, malondialdehyde; miRNA, mitochondrial Ribonucleic acid; MMP, matrix metalloproteinase; NF-κB, nuclear factor kappa B; ROS, reactive oxygen species; SOD, superoxide dismutase; TPL, Triphala; XO, xanthine oxidase.

It may be in part modulated the radiation sensitive enzymes/radiation centers, hence protected cells from radiation induced damage (Sandhya et al., 2006). Additionally, Sharma et al. (2011) showed that ellagic acid (EA) prevented endoplasmic stress mediated peroxidative damage by upregulating glutathione and also diverted the active metabolites (electrophiles) of toxin dimethyl hydrazine dihydrochloride (DMH) from their interactions with critical cellular bio-molecules which could be responsible for its protective action against DMH (Sharma et al., 2011).

Furthermore, Cao and co-workers (1998) showed that the consumption of controlled diet rich in fruits and vegetables (strawberry and spinach) increased significantly the antioxidants capacity of plasma (Cao et al., 1998). Additionally, epidemiological studies have found that there is a significant negative association between the intake of fruits and vegetables and heart disease mortality (Hertog et al., 1993, Knekt et al., 1996). Consumption of fruits, vegetables, and whole grains may be a good strategy to increase antioxidant intake (Chun et al., 2010; Yang, 2013). Matejuk and Shamsuddin reported radioprotective property of two common dietary molecules found in legumes and bran (inositol and IP6) could protect DNA from the harmful effects of radiation (Matejuk and Shamsuddin, 2010). Furthermore, Leskovac and co-workers demonstrated dietary herbals can help in prevention of the injuries caused by ionizing irradiation and, therefore, they encourages studies on radioprotective properties of dietary foods (Leskovac et al., 2011).

6.3. Mechanism of Action

AOs exhibit their protective role by a number of ways such as electron donation, metal ion chelation, co-antioxidants, or regulation of gene expression (Krinsky, 1992). Researchers around the world believe that the protective activity of AOs broadly falls in two main categories. The first one is a chain- breaking mechanism by which the AOs donates electron to the ROS present in the systems. The other mechanism is quenching chain-initiating catalyst that involves removal of ROS inducers (secondary antioxidants) (Rice-Evans et al., 1993; Brewer, 2011). ROS scavenging by antioxidants interferes with the initiation of apoptosis with down-regulation of ROS levels in healthy cells and maintaining membrane integrity (Salganik, 2001; Denissova et al., 2012; Russo et al., 2012). Antioxidants also have the ability to affect apoptosis through inhibiting proteins in the apoptotic cascade or modification of gene expression. By inhibiting the cleavage of caspase-3 and its substrate, poly(ADP-ribose) polymerase, the green tea polyphenol, (−)-epigallocatechin, was found to prevent apoptosis in HaCaT human keratinocytes (Kondo et al., 2004).

These herbal antioxidants have potential to modify cell survival and DNA repair efficacy by down-regulating radiation-induced activation of receptor tyrosine kinases and nuclear factor-κB signaling (Nambiar et al., 2011). Since, plants and herbs have a complex mixture of many chemicals therefore, multiple mechanism can act simultaneously on varying cellular environment within the host (Vrinda et al., 2001; Sandhya et. al, 2006; Sharma et al., 2011). Up-regulation of antioxidants like CAT, GSH counteract to ROS production via irradiation. Furthermore, up-regulation of DNA repair genes initiate error free repair of damaged DNA. The molecular mechanism of plants inhibits regulatory factors such as Protein Kinase c (PKC), MAPK, cyt.P450, JNK, nitric oxide and genes responsible for inducing DNA damage (Katiyar et al., 2001)

6.4. Pro- and Anti-Apoptotic Factors

From the voluminous experimental evidences we perceive that indigenous dietary plants possess a broad spectrum of bioactive compounds such as polyphenols, flavonoids, alkaloids, tannins, allicin; that have the potential to protect normal cells/tissues from harmful effects of ionizing radiation. The molecular mechanism is reported to involve differences in their ability to induce apoptosis in stressed cells and inhibit in normal tissues. It seems that AOs recognize the cytosolic redox status of cells which is known to be different in normal and stressed cells, hence accordingly behave as pro-oxidant or anti-oxidant. Accordingly, the pro-apoptotic factors are those that sensitize the cell towards radiation induced damage. ROS inducing/up-regulating factors, mutated p53, cell cycle phase, Enzyme Receptor Kinases. Anti-apoptotic factors are ROS scavenging that inhibits apoptosis, offer resistance towards radiation, subsequently protecting them during and after irradiation. ROS scavenging by antioxidants interferes with apoptosis initiation by down-regulating ROS levels in cells, thus maintains membrane integrity. The most important factor that ensures cell survival is release of transcription factor NF-κB. Cell cycle checkpoints, are also important factors determining sensitivity or resistance towards radiation exposure and DNA damage. They are responsible for delay in cell cycle progression in order to provide more time for repair.

In vitro and *in vivo* investigations of herbal formulation Triphala, ginger, Indian berries and many other dietary herbals provide evidence for apoptosis of diseased cell by ROS dependent apoptotic mechanism (Sandhya, Lathika et al., 2006; Jagetia et al., 2003; Rao et al., 2011; Mishra and Bhataia, 2010). The status of p53 in tumor cells is another important factor in predicting the response of tumor cells to pro-oxidant drugs(Sandhya and Mishra, 2006). Experimental work of Y Shi and coworkers(2008) supports the fact that Triphala failed to induce apoptosis or activate ERK or p53 in normal human pancreatic ductal epithelial (HPDE-6) cells, however induced apoptosis with phosphorylation of p53 at Ser-15 and ERK at Thr-202/Tyr-204 in Capan-2 cells (Shi et al., 2008). Moreover, oral administration of Triphala significantly suppresses the growth of pancreatic tumor xenograft in nude mice by inducing apoptosis and activation of ERK and p53 in the tumor (Shi et al., 2008).

7. Practical Approaches for Radiation Protection

The practical approach to introduce these potent herbal antioxidants into mainstream requires scientific validation involving various stages of translational research, dose optimization, systematic clinical trials (Sagar et al., 2008; Nambiar et al., 2011).

Potent antioxidants are needed that could improve the efficacy of radio-protection, prevent the damage to normal cells and tissues from the direct and/or indirect (bystander) effects of radiation; devoid of their own systemic toxicity. About three-fourth of the compounds used in chemotherapy such as taxol, vinca alkaloids are derived from herbal sources. Numerous dietary AOs and herbals are capable in providing significant protection against harmful radiations. Dietary phytochemicals (usually non-toxic agents with both radioprotective and therapeutic activities) could rightly fit in this approach. Therefore, their contribution in modulation of long-term radiation effects such as in cancer, needs further examination. Numerous, experimental evidences contribute towards low cost, easy and reliable therapeutic scope of herbal antioxidants as radio-protector. Although based on few reports claiming incapability of single antioxidants in providing protection against irradiation, researchers also suggested administration of multiple antioxidants such as vitamin A, beta-carotene, vitamin C and vitamin E (Prasad, 2005; Prasad and Cole, 2006). Pharmacological properties of herbs and their derivative enables them to distribute differentially or activate and to have selective metabolic effects among tumor cells and normal tissues (Wong et al., 2010). However, there remains some kind of controversy in these reports and basic question continues to pose a challenge and scope exists for further experimentation and general agreement by clinicians.

8. Perspective

Antioxidants of herbal origin have unique ability to discriminate between different cellular environments of cells on the basis of radiation centers and other altered cellular moieties such as mutated p53, up-regulated transcription factor hence, act accordingly (Sandhya and Mishra, 2006; Sandhya et al., 2006). Therefore, screening of such herbals is more advisable that protect normal cells from harmful effects of radiation and enhance radiation mediated apoptosis of tumor cells, hence contribute in safe cancer radiotherapy. Moreover, after reviewing the significant and promising role of dietary antioxidants against harmful effects of radiation and related consequences of stress, we assume that herbals may also be quite a help in reducing the risk related to radiation exposure during nuclear disasters and radiation-induced oxidative stress in space flights too. Inclusion of such dietary herbals into food habits might be helpful in for people of medical care, space research, those residing/working near nuclear plants.

However, very little is known regarding the radioprotective effects of antioxidants on long term radiation effects, especially for antioxidants with dietary/herbal origin, there is only a limited understanding of these effects at a clinical level. Therefore, additional studies requisites to explore these current and new herbal antioxidant compounds for their radioprotective effects in cells and tissues exposed to radiation. This will definitely support rational approaches projecting worthy dietary antioxidants as radioprotector.

Conclusion

Based on results of many scientific findings we are inclined to believe that antioxidants have potential to reduce the consequences of radiation exposure in normal cells. There is a need to look for a compound that can act either as pro-apoptotic or anti-apoptotic according to cytosolic status of the cell. Under normal cellular environment it should offer resistance to IR, neutralize/scavenge ROS mediated oxidative stress. On the other hand, it should be capable enough to sensitize macromolecules of infected /diseased cells towards IR hence, promote apoptosis. Several experimental reports have appeared including the results from authors' laboratory that dietary AOs such as triphala, ellagic acid and eugenol substantially protected normal cells or at least not affected normal cells during radiation exposure. It however remains to be extensively investigated in animal experiments and human trials.

The long-term consequences of radiation toxicity varies but may be greatly alleviated by slight /healthy modifications in food habits. We suggest including balanced and consistent antioxidant rich dietary intakes that in long term may potentially yield a preventative/therapeutic strategy to prevent adverse effects from both short and long term radiation exposure.

Apart from established radioprotectors, indigenous plants and herbal formulations including wide range of dietary herbals encompassing potent radioprotector antioxidants yet need to be explored further for their scientific validation.

REFERENCES

[1] Aiyer, H. S., Vadhanam, M. V., Stoyanova, R., Caprio, G. D., Clapper, M. L., & Gupta, R. C. (2008). Dietary berries and ellagic acid prevent oxidative DNA damage and modulate expression of DNA repair genes. *International journal of molecular sciences*, 9(3), 327-341.

[2] Akyol, S., Ginis, Z., Armutcu, F., Ozturk, G., Yigitoglu, M. R., & Akyol, O. (2012). The potential usage of caffeic acid phenethyl ester (CAPE) against chemotherapy-induced and radiotherapy-induced toxicity. *Cell biochemistry and function*, *30*(5), 438-443.

[3] Alvarez, J. G., & Aitken, R. J. (2012). Lipid Peroxidation in Human Spermatozoa", *Studies on men's health and fertility*, 119-130.

[4] Averbeck, D. (2010). Non-targeted effects as a paradigm breaking evidence. *Mutation Research/Fundamental and Molecular Mechanisms of Mutagenesis*, *687*(1), 7-12.

[5] Belli, M., Sapora, O., & Tabocchini, M. A. (2002). Molecular targets in cellular response to ionizing radiation and implications in space radiation protection. *Journal of radiation research*, 43(Suppl), S13-S19.

[6] Bravard, C. Luccioni, E. Moustacchi, O. Rigaud, A. (1999). Contribution of antioxidant enzymes to the adaptive response to ionizing radiation of human lymphoblasts. *International journal of radiation biology*,*75*(5), 639-645.

[7] Brewer, M. S. (2011). Natural antioxidants: sources, compounds, mechanisms of action, and potential applications. Comprehensive Reviews in Food Science and Food Safety, 10(4), 221-247.

[8] Brizel, D. M., Wasserman, T. H., Henke, M., Strnad, V., Rudat, V., Monnier, A., ...& Sauer, R. (2000). Phase III randomized trial of amifostine as a radioprotector in head and neck cancer. *Journal of Clinical Oncology*, 18(19), 3339-3345.

[9] Burdon, R. H. (1996). Control of cell proliferation by reactive oxygen species. *Biochemical Society Transactions*, 24(4), 1028-1032.

[10] Cai, F., Dupertuis, Y. M., & Pichard, C.(2012). Role of polyunsaturated fatty acids and lipid peroxidation on colorectal cancer risk and treatments. Current opinion in *clinical nutrition & metabolic* care, 15(2), 99.

[11] Capizzi, R. L., & Oster, W. (2000). Chemoprotective and radioprotective effects of amifostine: an update of clinical trials. *International journal of hematology*, *72*(4), 425.

[12] Cao, G., Russell, R. M., Lischner, N., & Prior, R. L. (1998). Serum antioxidant capacity is increased by consumption of strawberries, spinach, red wine or vitamin C in elderly women. *The Journal of nutrition, 128*(12), 2383-2390.

[13] Chun, O. K., Floegel, A., Chung, S. J., Chung, C. E., Song, W. O., & Koo, S. I. (2010). Estimation of antioxidant intakes from diet and supplements in US adults. *The Journal of nutrition, 140*(2), 317-324.

[14] Cho, J. W., Park, K., Kweon, G. R., Jang, B. C., Baek, W. K., Suh, M. H., ... & Suh, S. I. (2005). Curcumin inhibits the expression of COX-2 in UVB-irradiated human keratinocytes (HaCaT) by inhibiting activation of AP-1: p38 MAP kinase and JNK as potential upstream targets. *Exp. Mol. Med., 37*(3), 186-92.

[15] Collis, S. J., Schwaninger, J. M., Ntambi, A. J., Keller, T. W., Nelson, W. G., Dillehay, L. E., & DeWeese, T. L. (2004). Evasion of early cellular response mechanisms following low level radiation-induced DNA damage. *Journal of Biological Chemistry, 279*(48), 49624-49632.

[16] Denissova, N. G., Nasello, C. M., Yeung, P. L., Tischfield, J. A., & Brenneman, M. A. (2012). Resveratrol protects mouse embryonic stem cells from ionizing radiation by accelerating recovery from DNA strand breakage. *Carcinogenesis,33*(1), 149-155.

[17] Eken, A., Aydin, A., Erdem, O., Akay, C., Sayal, A., & Somuncu, I. (2012). Induced antioxidant activity in hospital staff occupationally exposed to ionizing radiation. *International journal of radiation biology, 88*(9), 648-653.

[18] Giardi, M.T., Touloupakis, E., Bertolotto D.& Mascetti G. (2013). Preventive or Potential Therapeutic Value of Nutraceuticals against Ionizing Radiation-Induced Oxidative Stress in Exposed Subjects and Frequent Fliers, *Int. J. Mol. Sci.* Aug 2013; 14(8): 17168–17192.

[19] Grdina, D. J., Murley, J. S., Kataoka, Y., Baker, K. L., Kunnavakkam, R., Coleman, M. C., & Spitz, D. R. (2009). Amifostine induces antioxidant enzymatic activities in normal tissues and a transplantable tumor that can affect radiation response. *International Journal of Radiation Oncology* Biology* Physics*, 73(3), 886-896.

[20] Güngör, N., Knaapen, A. M., Munnia, A., Peluso, M., Haenen, G. R., Chiu, R. K., ...& Van Schooten, F. J. (2010). Genotoxic effects of neutrophils and hypochlorous acid. *Mutagenesis*;25(2), 149-154.

[21] Hamanaka, R. B., & Chandel, N. S., "Mitochondrial reactive oxygen species regulate cellular signaling and dictate biological outcomes", *Trends in biochemical sciences* (2010), 35(9), 505-513.

[22] Halliwell B., Gutteridge J. M. C. (1989). *Free Radicals in Biology and Medicine*, 2nd Edn. Oxford: Clarendon Press.

[23] Hertog, M. G., Feskens, E. J., Kromhout, D., Hertog, M. G. L., Hollman, P. C. H., Hertog, M. G. L., & Katan, M. B. (1993). Dietary antioxidant flavonoids and risk of coronary heart disease: the Zutphen Elderly Study. *The Lancet,342*(8878), 1007-1011.

[24] Jagetia, G.C., Baliga, M.S., Venkatesh, P., and Ulloor, J.N. (2003), Influence of ginger rhizome (*Zingiber officinale* Rosc) on survival, glutathione and lipid peroxidation in mice after whole-body exposure to gamma radiation. *Radiat Res.*, 160, 584-592.

[25] Janssen, Y. M., Van Houten, B., Borm, P. J., & Mossman, B. T. (1993). Cell and tissue responses to oxidative damage. *Laboratory investigation; a journal of technical methods and pathology, 69*(3), 261-274.

[26] Kamata, H., & Hirata, H. (1999). Redox regulation of cellular signalling. *Cellular signalling*, *11*(1), 1-14.

[27] Kemp, G., Rose, P., Lurain, J., Berman, M., Manetta, A., Roullet, B., ...& Glick, J. (1996). Amifostine pretreatment for protection against cyclophosphamide-induced and cisplatin-induced toxicities: results of a randomized control trial in patients with advanced ovarian cancer. *Journal of clinical oncology*, 14(7), 2101-2112.

[28] Khafif, A. V. I., Hurst, R., Kyker, K., Fliss, D. M., Gil, Z. I. V., & Medina, J. E. (2005). Curcumin: a new radio-sensitizer of squamous cell carcinoma cells. *Otolaryngology-Head and Neck Surgery*, 132(2), 317-321.

[29] Kim, W. H., Park, W. B., Gao, B., & Jung, M. H. (2004). Critical role of reactive oxygen species and mitochondrial membrane potential in Korean mistletoe lectin-induced apoptosis in human hepatocarcinoma cells. *Molecular pharmacology*, *66*(6), 1383-1396.

[30] Kim, J. K., Kim, Y., Na, K. M., Surh, Y. J., & Kim, T. Y. (2007). [6]-Gingerol prevents UVB-induced ROS production and COX-2 expression in vitro and in vivo. *Free radical research*, *41*(5), 603-614.

[31] Knekt, P., Jarvinen, R., Reunanen, A., & Maatela, J. (1996). Flavonoid intake and coronary mortality in Finland: a cohort study. *BMJ: British Medical Journal*,*312*(7029), 478.

[32] Kondo H, Park S-H, Watanabe K, et al.(2004). Polyphenol (−)-epigallocatechin gallate inhibits apoptosis induced by irradiation in human HaCaT keratinocytes. *Biochem. Biophys. Res. Commun*;316:59–64.

[33] Kouvaris, J. R., Kouloulias, V. E., & Vlahos, L. J. (2007). Amifostine: the first selective-target and broad-spectrum radioprotector. *The oncologist*, 12(6), 738-747.

[34] Krinsky, N. I. (1992). Mechanism of action of biological antioxidants. *Experimental Biology and Medicine*, *200*(2), 248-254.

[35] Limón-Pacheco, J., & Gonsebatt, M. E. (2009). The role of antioxidants and antioxidant-related enzymes in protective responses to environmentally induced oxidative stress. *Mutation Research/Genetic Toxicology and Environmental Mutagenesis*, *674*(1), 137-147.

[36] Liu, Q., Jiang, B., Jiang, L. P., Wu, Y., Wang, X. G., Zhao, F. L., ... & Jiang, E. (2008). Clinical report of three cases of acute radiation sickness from a 60Co radiation accident in Henan province in China. *Journal of radiation research*,*49*(1), 63-69.

[37] Lobo, V., Patil, A., Phatak, A., & Chandra, N. (2010). Free radicals, antioxidants and functional foods: Impact on human health. *Pharmacognosy reviews*, *4*(8), 118.

[38] Manda Gina, Nechifor Marina Tamara and Neagu Teodora-Monica (2009),"Reactive Oxygen Species, Cancer and Anti-Cancer Therapies", *Current Chemical Biology, 3, 342-366.*

[39] Mansour, H. H., & Tawfik, S. S. (2012). Early treatment of radiation-induced heart damage in rats by caffeic acid phenethyl ester. *European journal of pharmacology*, *692*(1), 46-51.

[40] Matejuk, A., & Shamsuddin, A. (2010). IP6 in cancer therapy: past, present and future. *Current Cancer Therapy Reviews*, *6*(1), 1-12.

[41] McDonald, J. T., Kim, K., Norris, A. J., Vlashi, E., Phillips, T. M., Lagadec, C., ... & McBride, W. H. (2010). Ionizing radiation activates the Nrf2 antioxidant response. *Cancer research*, *70*(21), 8886-8895.

[42] Mettler, F. A. (2012). Medical effects and risks of exposure to ionising radiation. *Journal of Radiological Protection*, *32*(1), N9.

[43] Mishra, T., & Bhatia, A. (2010). Augmentation of expression of immunocytes' functions by seed extract of Ziziphus mauritiana (Lamk.). *Journal of ethnopharmacology*, 127(2), 341-345.

[44] Nambiar, D., Rajamani, P., & Singh, R. P. (2011). Effects of phytochemicals on ionization radiation-mediated carcinogenesis and cancer therapy. *Mutation Research/Reviews in Mutation Research*, *728*(3), 139-157.

[45] Narang, H., Bhat, N., Gupta, S. K., Santra, S., Choudhary, R. K., Kailash, S., & Krishna, M. (2009). Differential activation of mitogen-activated protein kinases following high and low LET radiation in murine macrophage cell line. *Molecular and cellular biochemistry*, *324*(1-2), 85-91.

[46] Narici, L., De Martino, A., Brunetti, V., Rinaldi, A., Sannita, W. G., & Paci, M. (2009). Radicals excess in the retina: A model for light flashes in space. *Radiation Measurements*, *44*(2), 203-205.

[47] Okunieff, P., Swarts, S., Keng, P., Sun, W., Wang, W., Kim, J., ... & Zhang, L. (2008). Antioxidants reduce consequences of radiation exposure. In *Oxygen Transport to Tissue XXIX* (pp. 165-178). Springer US.

[48] Ott, M., Gogvadze, V., Orrenius, S., & Zhivotovsky, B. (2007). Mitochondria, oxidative stress and cell death. *Apoptosis*, *12*(5), 913-922.

[49] Pandey, B. N., & Mishra, K. P. (2004). Modification of thymocytes membrane radiooxidative damage and apoptosis by eugenol. *Journal of environmental pathology, toxicology and oncology*, 23(2).

[50] Pandey, B. N., Lathika, K. M., & Mishra, K. P. (2006). Modification of radiation-induced oxidative damage in liposomal and microsomal membrane by eugenol. *Radiation Physics and Chemistry*, 75(3), 384-391.

[51] Pandey, B. N., Kumar, A., Ali, M., & Mishra, K. P. (2011). Bystander effect of conditioned medium from low and high doses of γ-irradiated human leukemic cells on normal lymphocytes and cancer cells. *Journal of Environmental Pathology, Toxicology and Oncology*, *30*(4).

[52] Patt, H. M., Tyree, E. B., Straube, R. L., & Smith, D. E. (1949). Cysteine protection against X irradiation. *Science* (New York, NY), 110(2852), 213.

[53] Prasad, K. N. (2005). Rationale for using multiple antioxidants in protecting humans against low doses of ionizing radiation. *Br. J. Radiol.* 78, 485–492.

[54] Prasad, K. N., and Cole, W. C. (2006). Antioxidants in cancer therapy. *J. Clin. Oncol.* 24, e8–e9.

[55] Rao, B. N., Archana, P. R., Aithal, B. K., & Rao, B. S. S. (2011). Protective effect of zingerone, a dietary compound against radiation induced genetic damage and apoptosis in human lymphocytes. *European journal of pharmacology*, 657(1), 59-66.

[56] Rastogi, L., Feroz, S., Pandey, B. N., Jagtap, A., & Mishra, K. P. (2010). Protection against radiation-induced oxidative damage by an ethanolic extract of Nigella sativa L. *International Journal of Radiation Biology*, 86(9), 719-731.

[57] Rades, D., Fehlauer, F., Bajrovic, A., Mahlmann, B., Richter, E., & Alberti, W. (2004). Serious adverse effects of amifostine during radiotherapy in head and neck cancer patients. *Radiotherapy and oncology*, *70*(3), 261-264.

[58] Rice-Evans, C. A., & Diplock, A. T. (1993). Current status of antioxidant therapy. *Free Radical Biology and Medicine*, *15*(1), 77-96.

[59] Riley, P. A. (1994). Free radicals in biology: oxidative stress and the effects of ionizing radiation. *International journal of radiation biology*, *65*(1), 27-33.

[60] Ruch W, Cooper PH, Baggiolini M.(1983). Assay of H2O2 production by macrophages and neutrophils with homovanillic acid and horse-radish peroxidase". *J. Immunol. Methods*, 63:347–357.

[61] Russo, G. L., Tedesco, I., Russo, M., Cioppa, A., Andreassi, M. G., & Picano, E. (2012). Cellular adaptive response to chronic radiation exposure in interventional cardiologists. *European heart journal*, *33*(3), 408-414.

[62] Sagar, S. M., & Wong, R. K. (2008). Chinese medicine and biomodulation in cancer patients—Part one. *Current Oncology*, *15*(1), 42.

[63] Salganik RI (2001). The benefits and hazards of antioxidants: controlling apoptosis and other protective mechanisms in cancer patients and the human population. *J. Am. Coll. Nutr*;20:464S–472S.

[64] Sandhya, T., Lathika, K. M., Pandey, B. N., & Mishra, K. P. (2006). Potential of traditional ayurvedic formulation, Triphala, as a novel anticancer drug. *Cancer letters*, *231*(2), 206-214.

[65] Sandhya, T., Lathika, K. M., Pandey, B. N., Bhilwade, H. N., Chaubey, R. C., Priyadarsini, K. I., & Mishra, K. P. (2006). Protection against radiation oxidative damage in mice by Triphala. *Mutation ResearchGenetic Toxicology and Environmental Mutagenesis*, 609(1), 17-25.

[66] Sandhya, T., & Mishra, K. P. (2006). Cytotoxic response of breast cancer cell lines, MCF 7 and T 47 D to triphala and its modification by antioxidants. *Cancer letters*, *238*(2), 304-313.

[67] Sharma, A., & Sharma, K. K. (2011). Chemoprotective role of triphala against 1, 2-dimethylhydrazine dihydrochloride induced carcinogenic damage to mouse liver. *Indian Journal of Clinical Biochemistry*, 26(3), 290-295.

[68] Shi, Y., Sahu, R. P., & Srivastava, S. K. (2008). Triphala inhibits both in vitro and in vivo xenograft growth of pancreatic tumor cells by inducing apoptosis. *BMC cancer*, 8(1), 294.

[69] Shimizu, M., Deguchi, A., Lim, J. T., Moriwaki, H., Kopelovich, L., & Weinstein, I. B. (2005). (−)-Epigallocatechin gallate and polyphenon E inhibit growth and activation of the epidermal growth factor receptor and human epidermal growth factor receptor-2 signaling pathways in human colon cancer cells. *Clinical Cancer Research*, *11*(7), 2735-2746.

[70] Sinclair, W. K. (1968). Cysteamine: differential x-ray protective effect on Chinese hamster cells during the cell cycle. *Science*, *159*(3813), 442-444.

[71] Singh, S. P., Abraham, S. K., & Kesavan, P. C. (1996). Radioprotection of mice following garlic pretreatment. The British journal of cancer. Supplement, 27, S102.

[72] Spitz, D. R., Azzam, E. I., Li, J. J., & Gius, D. (2004). Metabolic oxidation/reduction reactions and cellular responses to ionizing radiation: a unifying concept in stress response biology. *Cancer and Metastasis Reviews*,*23*(3-4), 311-322.

[73] Stadtman, E. R. (2004). Role of oxidant species in aging. *Current medicinal chemistry*, *11*(9), 1105-1112.

[74] Valko, M., Leibfritz, D., Moncol, J., Cronin, M. T., Mazur, M., & Telser, J. (2007). Free radicals and antioxidants in normal physiological functions and human disease. *The international journal of biochemistry & cell biology*, *39*(1), 44-84.

[75] Velho-Pereira, R., Kumar, A., Pandey, B. N., Jagtap, A. G., & Mishra, K. P. (2011). Radiosensitization in human breast carcinoma cells by thymoquinone: role of cell cycle and apoptosis. Cell biology international, 35(10), 1025-1029.

[76] Vrinda, B., & Uma Devi, P. (2001). Radiation protection of human lymphocyte chromosomes in vitro by orientin and vicenin. *Mutation Research/Genetic Toxicology and Environmental Mutagenesis*, *498*(1), 39-46.

[77] Yang, M., Chung, S. J., Floegel, A., Song, W. O., Koo, S. I., & Chun, O. K. (2013). Dietary antioxidant capacity is associated with improved serum antioxidant status and decreased serum C-reactive protein and plasma homocysteine concentrations. *European journal of nutrition*, *52*(8), 1901-1911.

[78] Zabbarova, I., & Kanai, A. (2008). Targeted delivery of radioprotective agents to mitochondria. *Molecular Interventions*, 8(6), 294.

[79] Zong, H., Wang, F., Fan, Q. X., & Wang, L. X. (2012). Curcumin inhibits metastatic progression of breast cancer cell through suppression of urokinase-type plasminogen activator by NF-kappa B signaling pathways. *Molecular Biology Reports*, *39*(4), 4803-4808.

In: Biological Responses, Monitoring and Protection … ISBN: 978-1-63463-852-4
Editor: K. P. Mishra

Chapter 14

ROLE OF TOPOISOMERASE IB ON DAMAGE INDUCED BY IONIZING RADIATION

Zeenat Jahan[1], Marina Cerreto[1], Giorgia Aversa[2], Stefano Rufini[1], Alessandro Desideri[1] and Anna Giovanetti[2]
[1]Department of Biology, University of Rome "Tor Vergata", Rome, Italy
[2]UTBIORAD, ENEA Casaccia, Rome, Italy.

ABSTRACT

Ionizing radiation (IR) can cause damage to DNA molecules, both through direct ionization and through induction of oxidative stress. Thus the resistance and the rescue to radiation-induced damage are mostly associated with the efficacy of DNA repair system, which involves different types of enzymes and signal molecules, many of which are not yet identified. These enzymes represent target for molecules that, modulating their activity, may increase or decrease the efficiency of radiation-induced DNA repair and consequently cell viability. Topoisomerases (Tops) are ubiquitous and essential enzymes that solve the DNA topological problems deriving from replication, transcription, chromatin assembly, recombination, and chromosome segregation by introducing transient breaks into the helix. Human Topoisomerase IB (Top IB) is also of significant medical interest being the selective target of antitumor drugs such as camptothecin and its water-soluble derivatives, widely used for treating cancers sometimes in association with radiotherapy. There are several evidences that Top IB is involved in DNA repair, although contradicting results have been reported. The role of Top IB in repairing DNA damage induced by X rays was studied on two human cancer cell lines (i.e. colorectal carcinoma - HCT 116 - and adenocarcinoma - HT 29) evaluating the relation between Top IB activity and the level of DNA damage. After IR exposure, the DNA repair is correlated to an increase of Top IB activity but not of its synthesis. Based on these outcomes and on the available literature, we figure that the radiation exposure induces a post-translational modification of the enzyme modulating its activity. Our future prospective is to identify these post-translational modifications, and also to find out relation between Top IB and other repair proteins. These results provide a new insight into the mechanisms for genomic integrity maintenance, and could also contribute to new strategies in cancer treatment.

Introduction

Ionizing radiation (IR) can cause damage to cellular structures, both through direct ionization and through production of reactive oxygen species that ultimately results in damage. Characteristic effects of IR are DNA single strand breaks (SSB) and double-strand breaks (DSB) (Hall and Giaccia, 2006). SSB are not directly cytotoxic but during DNA replication may generate potentially lethal DSB by collapse of stalled replication forks (Saleh-Gohari et al., 2005). Radiation-induced SSB are primarily fixed by base excision repair mechanisms (Caldecott, 2003) and other repair pathways most of them are not yet been identified. The outcome of early and long-term radiation-induced damage is mostly associated with the efficacy of DNA repair system, which involves different types of signal molecules and enzymes. These proteins represent target for molecules that, modulating their activity, may increase or decrease the efficiency of radiation-induced DNA repair and consequently effects.

Topoisomerases (Tops) are ubiquitous and essential enzymes that solve the DNA topological problems deriving from replication, transcription, chromatin assembly, recombination, and chromosome segregation by introducing transient breaks into the helix (Stewart et al., 1998). Tops are grouped into type I and type II. Type I Tops, usually monomeric, cleave one strand of duplex DNA and are further classified into two subfamilies: type IA and type IB. There are several evidences that Top IB is involved in DNA repair, although contradicting results have been reported (Synder et al., 1987). A key function of Top IB enzyme is relaxation of DNA supercoiling in the absence of an energy cofactor by nicking the DNA and enabling the broken strand to rotate around the Top IB-bound DNA strand. Therefore, DNA Top IB is particularly vulnerable to a group of inhibitors such as camptothecin (CPT) and its water-soluble derivatives: topotecan and irinotecan during their cleavage intermediate step (referred to as the cleavage complex) (Pommier, 2006). Hence Top IB rises significant medical interest being the selective target of CPT and its water-soluble derivatives, widely used for treating cancers some times in association with radiotherapy (Allan et al., 2004).

The level of Top IB can be highly variable amongst tumor specimens and cell lines and this can arise to differential cellular responses to CPT and related drugs (Reinhold et al., 2010; Padhyay et al., 2007; Pommier et al., 1999; Sugimoto et al., 1990). In addition, it is also apparent that cancer cells have mechanisms able to regulate Top IB activity in the absence of changes in protein expression (Husain et al., 1994; St-Amant et al., 2006). These mechanisms, such as phosphorylation status, have not been well delineated, although they may play an equal or major role in the clinical response to therapy. In OVCAR3 ovarian cancer cells, for example, the failure of ectopic overexpression of Top IB to increase overall Top IB activity and cellular sensitivity to CPT can be attributed to a reduced ability of this cell line to phosphorylate the enzyme (St-Amant et al., 2006). Tops were demonstrated to be a substrate of different kinases, including casein kinase 2, that is frequently up regulated in cancer cell lines (Olsen et al., 2012; Bandyopadhyay and Gjerset, 2011) and that is activated by the DNA breaks induced by different agents such as UV and free radicals (Olsen et al., 2012). Thus, targeting the post-translational modification of Top IB, in particular the phosphorylation pathway could explore a new pharmacological cancer's treatment. Top I B-targeted CPT derivatives represent a class of anticancer agents that exhibit selective radiosensitizing effect toward a variety of solid tumors (Chen et al., 2004) having, however, severe side effects (Chen et al., 1994; Liehr et al., 2000). The possibility to modulate Top IB activity by acting on its phosphorylation may reduce the use of Top I inhibiting drugs proven to have severe toxic effects, as myeolosuppresion (Ewesuedo, 1997). Top IB phosphorylation, in fact, can both reactivate the DNA relaxing activity and restore its susceptibility to CPT (Pfister et al., 2009). A better understanding of how Top IB activity is regulated is critical not only for the clinical application but also for improving our knowledge on the IR-induced DNA damage repair. Previous work on *Saccharomyces cerevisiae* showed that Top IB after an early higher amount of SSB, increased the repair rate of IR-induced DNA breaks. Accordingly early after irradiation Top IB activity showed a two-three times increasing, although the total concentration remains unchanged, suggesting a IR-induced post-translational modification (Jahan et al., 2013).

To get insight the role of Top IB in repairing IR effect, we studied the relationship between Top IB activity and DNA damage induced by X rays, on two human epithelial colorectal cell lines (HCT 116 and HT 29) showing, on the NCI-60 cancer cell line panel, the highest Tops expression level (Pfister et al., 2009).

MATERIALS AND METHODS

Chemicals

Dulbecco's modified Eagles's medium (DMEM), fetal bovine serum, penicillin, streptomycin, L-glutamine are provided from Biowhittaker (Rome, Italy); complete protease inhibitor cocktail from Roche (Mannheim Germany); goat anti-rabbit-HRP and goat anti-mouse-HRP from Biorad (Milano, Italy); ECL Western Blot Detection system from Amersham (Milano, Italy); mounting medium from Electron Microscopy Sciences (Washington, PA, USA). All the other chemicals were provided from Sigma-Aldrich (Milano Italy).

Cell Culture

Cancer cell lines (human epithelial colorectal carcinoma HCT 116 and human epithelial colorectal adenocarcinoma HT29) purchased from ATCC (Milano, Italy), were grown in DMEM supplemented with 10% fetal bovine serum, 100 units/ml penicillin, 100 units/ml streptomycin, and 1% L-glutamine in an atmosphere of 5% CO_2 and air at 37 °C.

Irradiation

Semi-confluent cells were irradiated at a dose-rate of 1.1 Gy/min X-rays generated by CHF 320 G (Gilardoni, Mandello del Lario, Italy) equipped with a Cu filter of 0.5 mm, operating at 250 KeV, 15 mA, for increasing intervals of time in order to obtain 0; 0.1; 1; 5; 15 Gy of X-rays.

Cell Cycle Analysis

24 hours following radiation exposure, cells were harvested, fixed in ice cold 96% ethanol and stored at -20°C. Before analysis cells were incubated for 15 min in 200 μg/ml RNase A at 37°C, and stained with 50 μg/ml propidium iodide. Cell cycle analysis was performed using a FACSCalibur flow cytometer (BD Biosciences, San Josè, CA) and the percentage of cells in preG1, G1, S, and G2/M phases was determined using the Cell Quest program (Itani et al., 2007).

Cytokinesis-block Micronucleus (CBMN) Assay

CBMN assay was performed according to the method described by Fenech (2010) with minor modifications. Cells seeded on coverslips and maintained in 6 wells-multiwell plates were irradiated and treated with cytochalasin B dissolved in dimethylosulfoxide (DMSO) to a final concentration of 3 μg/ml, 1 hour after. Twenty-four hours after cells were fixed with methanol: acetic acid (3:1) stored at RT until staining with ethydium bromide (10 μg/ml) and at least 500 binucleated (BN) cells for each experimental point were analyzed at 400 × magnification by a fluorescent Axiolab Zeiss microscope (Carl Zeiss AG, Oberkochen, Germany).

Comet Assay

One hour or 24 hour from irradiation cells were gently detached by rinsing three times with 37°C, Ca and Mg free, sterile PBS, and incubating at 37°C with 1 ml 0,25% trypsin/EDTA (T4049 Sigma) solution for around 4 min, from this step cells were maintained on ice under dim light conditions. The alkaline Comet assay was performed as described by Giovanetti et al. (2008). For microscopy analysis, slides were stained with ethidium bromide (10 μg/ml) immediately before being analyzed at 400 × magnification by a fluorescent Axiolab Zeiss microscope (Carl Zeiss AG, Oberkochen, Germany). Each experiment was repeated at least twice. For each slide a total of 100 cells was classified visually as belonging to one of five classes of damage from 0 (no visible tail) to 5 (most of DNA in tail) and expressed as arbitrary units (AU) (García et al., 2004). Visual scoring was demonstrated to have a clear relationship to the percentage of DNA appearing in the tail as measured with computer image analysis.

Preparation of Whole Cell Extract

Cells were washed with PBS, scraped in TEEG buffer (50mM Tris-HCl pH 7.5, 1mM EDTA, 1mM EGTA, 10% glycerol), and resuspend in Tris buffer (20mM Tris-HCl, pH 7.5) supplemented with inhibitors of protease (complete protease inhibitor cocktail) and phosphatase [phenylmethylsulfonyl fluoride (0.5 mM), DTT (1 mM), NaV (0.5 mM) and NaF

(5 mM)]. Cell lysis was performed by sonication (Branson, Danbury, CT) to obtain whole cell extract.

Preparation of Nuclear Protein Extract

Nuclear extract were prepared by a slightly modified procedure adapted from references (Mo et al., 2000; Hann et al., 1998). Briefly, cells were washed with PBS, scraped and collected into a micro-centrifuge tube. After a brief centrifugation to remove the supernatant, 1 ml cell lysis buffer [0.1% NP-40, mM Tris (pH 7.9), 10 mM $MgCl_2$, 15 mM NaCl] was added to the cell pellets and incubated on ice for 10 min. Nuclei were pelleted by centrifugation at 670 x g for 5 min, extracted in extraction buffer [50 mM NaCl, 20 mM Hepes (pH 7.9), 20% glycerol] for 1 hour at 4° C and then centrifuged at 5400 x g for 10 min to pellet the residual nuclear material. The supernatant fraction was termed nuclear extract. In all buffer was present the above reported proteases and phosphatases inhibitors cocktail.

Top IB Amount and Activity Quantification

Western analysis of Top IB amount in nuclear extract was carried out after 30 and 60 min from IR exposure using 50 μg of total protein. Protein concentration was determined by Lowry method (Lowry et al., 1951). Briefly, protein samples were separated by 13% SDS–PAGE and transferred to a nitrocellulose membrane in a semidry protein transfer apparatus. For detection of Top IB, we used a mouse monoclonal antibody recognizing a Top IB N-terminal epitope (BioRad, Hercules, CA) and as secondary antibodies goat anti-mouse-HRP. Both primary and secondary antibodies were used at 1:1000 and 1:10,000 dilutions for Western analysis. Signals were visualized with the ECL Western Blot Detection system. Relative intensity of protein bands in an X-ray film was determined by Fluorochem SP (Alpha Innotech, San Josè, CA) program.

For the activity assay, nuclear extract was prepared after 30 and 60 min from IR exposure, as described, followed by performing in-vitro Top IB activity by the method of Castelli et al. (2009). In brief, decreasing concentrations (5 μg to 0.6 μg) of nuclear extracts were incubated in 30 μl reaction volume containing 0.5 μg of negatively supercoiled pBlue Script KS II (+) DNA and reaction buffer (20 mM Tris-HCl, 0.1 mM Na_2EDTA, 5 mM $MgCl_2$, 50 μg/ml acetylated BSA and 150 mM KCl, pH:7.5) for 1 hour at 37 °C. Following stopping the reaction with a final concentration of 0.5 % SDS, samples were electrophoresed in a horizontal 1 % agarose gel (in 50 mM Tris, 45 mM boric acid, 1 mM EDTA). After staining with ethidium bromide (5μg/ml) and destaining with water the gel was photographed under UV illumination.

DNA-Top IB Binding Assay

For gel shift assay, after 10, 30 and 60 min from when IR cells were lysed in 1% N-lauroylsarcosine (pre-equilibrated to 37 ^{0}C) containing protease and phosphatase inhibitor.

Therefore, the resulting whole cell extracts were sonicated (Branson, Danbury, CT) to break down chromosomal DNA (Yang and Champoux, 2009) and protein concentrations were determined according to Lowry (Lowry et al., 1951). For immune-blotting, aliquots of crude cellular extract (50 μg) were separated by 10% SDS polyacrylamide gel, transferred to nitrocellulose membrane and blots were developed to detect only the DNA bound Top IB as previously described.

RESULTS

Effect of IR on Cell Cycle, Induction of MN and DNA Breaks

As Figure 1 shows in HT 29 cell the amount of cells in G2 increased with dose and treatment with 15 Gy induced the complete arrest of all cells in G2-phase. A similar response was also observed in HCT 116 cells.

HCT 116 cells showed a higher background level of MN (17/500 BN) comparing to HT 29 cells (10/500 BN). A dose-dependent increase of MN cells frequency has been observed for both cancer cell lines with R^2 values of 0.95 in HCT 116 and 0.98 in HT 29 cells (Figure 2).

Basal level of DNA breaks was also high (60.75 and 89 A.U. in HT 29 and HCT 116 cells) respect to no-cancer cell lines confirming that cancer genome is driven by genomic instability (Jenkins et al., 2013). One hour after IR exposure the amount of DNA breaks increased with dose in both cell lines. After 24 hours from IR, DNA damage in HT 29 cells decreased only in the cells exposed to 5 Gy, showing the higher degree of damage, while in HCT 116 cells DNA breaks' number sharply decreased in all dose-points (Figure 3).

Effect of IR on Top IB activity

The total amount of Top IB in cellular extracts didn't show any time or dose-related variation (Figure 4). To determine how the effect of IR correlates with endogenous levels of Top IB activity, we carried out in vitro Top IB activity assays (Figure 5) on different quantities of nuclear extracts (0.6 to 5 μg/ reaction) at different time points from IR exposure. The assays measure the ability of Top I to convert a supercoiled plasmid "s" to a slower migrating relaxed form "r" as revealed by agarose gel electrophoresis and ethidium bromide staining. As showed in Figure 4 we observed in HCT 116 cells a clear increasing in Top IB activity at 30 min from IR that decreased to the level of control at 60 min (Figure 5A), whereas in HT 29 cells activity remained constant at 30 min and increased at 60 min (Figure 5B). The ability of Top IB to bind the DNA molecule (Figure 5) following IR was evaluated through shift assay, which is based on the principle that Top IB bound to DNA does not enter into the acrylamide gel. Immunoblot analysis for the shift experiment performed in all dose-points at 10, 30 and 60 min from IR showed that at these times the amount of free Top IB in HCT 116 cells was not different from controls (Figure 5A), while HT 29 cells showed at all time points after IR a decrease of free Top IB (Figure 5B).

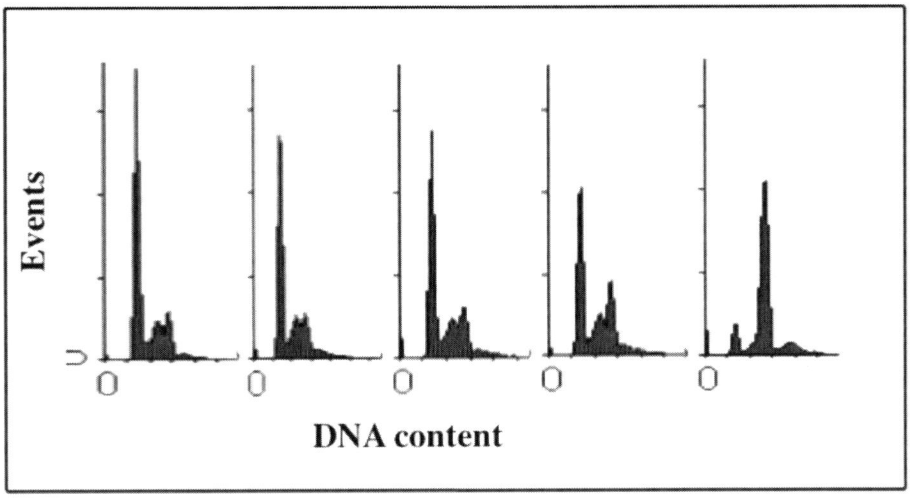

Figure 1. Cell cycle analysis of HT 29 cells irradiated with different doses of X rays. DNA content was analyzed by flow cytometry as described in Materials and Methods. Results are representative of three similar experiments with triplicates.

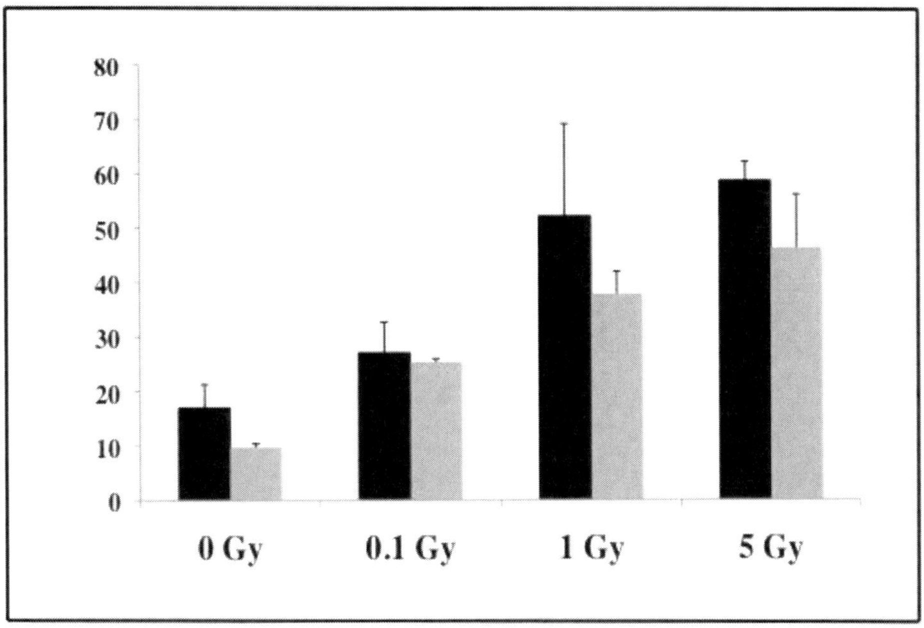

Figure 2. Amount of MN/500 BN cells in HCT 116 (black) and HT 29 (gray) cells. Each data point is representative of at least 3 individual experiments under same experimental conditions. Data represent the mean ± SEM of at least 3 independent experiments.

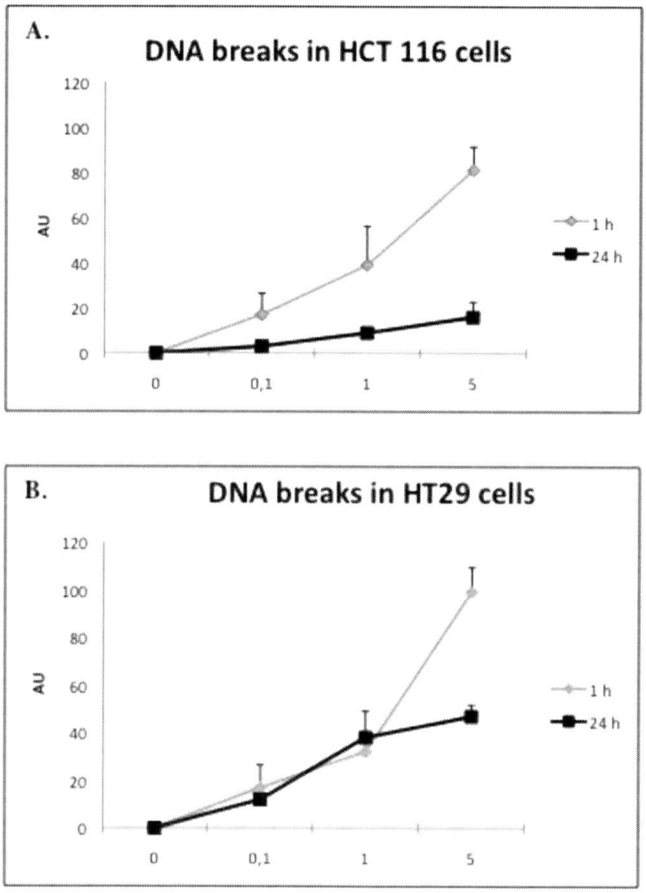

Figure 3. Comet assay analysis of the induction of DNA breaks in HCT 116 (A) and HT 29 (B) cell lines at different doses of IR, incubated for 1hour (grey line) or 24 hours (black line) after IR. Data are expressed as the increase in DNA breaks (measured as AU - arbitrary units) compared to non-irradiated cells of the same time point. Data represent the mean ± SEM of at least 3 independent experiments.

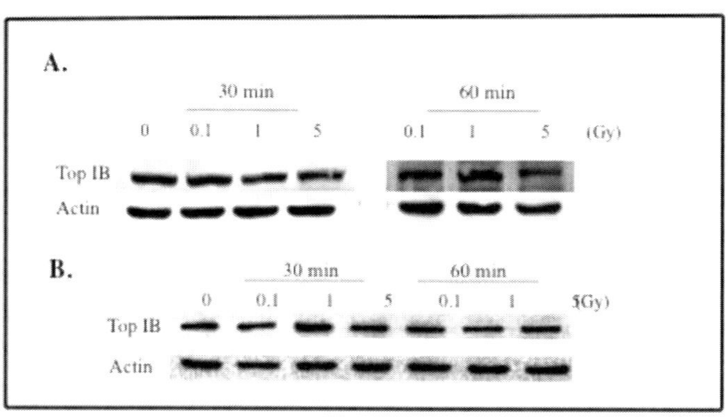

Figure 4. WB analysis of Top IB in extracts of HCT 116 (A) and HT 29 (B) in un-irradiated ('0') or irradiated cells (0.1, 1 and 5 Gy of X ray) after 30 or 60 min from IR exposure. Experiments were performed twice in triplicate, giving comparable results.

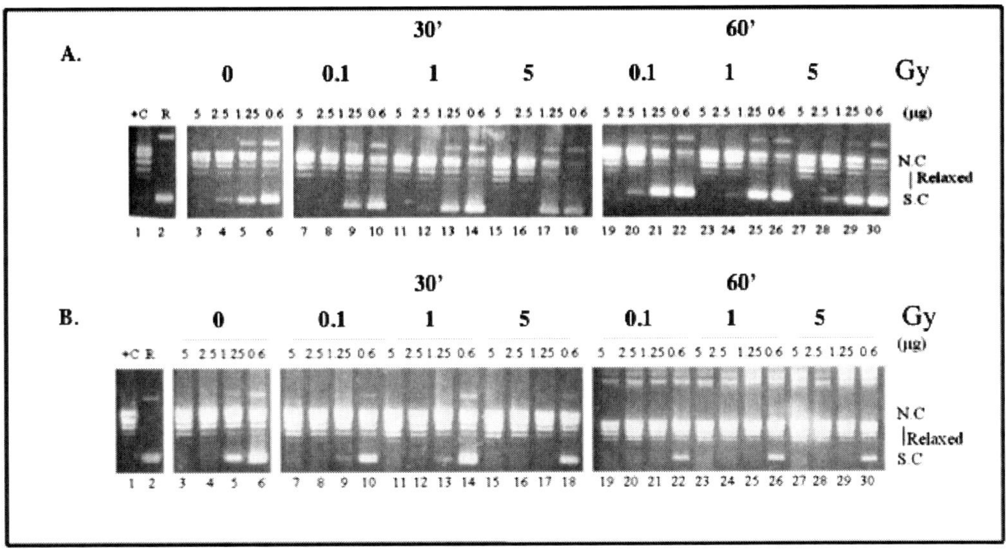

Figure 5. In vitro relaxation of negative supercoiled plasmid DNA by nuclear extract of (A) HCT 116 and (B) HT 29 cells, from not irradiated ('0') and irradiated with different doses of IR (0.1, 1 and 5 Gy). Lane 1, positive control reaction with DNA and purified Top IB. Lane 2, negative control reaction without enzyme. Lanes 3–6, activity of decreasing nuclear extract concentrations (5-0.6 µg) from control (without IR). Lanes 7–18, activity of nuclear extracts from cells 30 min following different doses (0.1, 1, and 5 Gy) of IR exposure using same series of dilution (5-0.6 µg) as control. Lanes 19-30, activity after 60 min of exposure using same doses (0.1, 1, and 5 Gy) of IR and dilutions (5-0.6 µg) of nuclear extracts. NC, nicked circular plasmid DNA. SC, supercoiled plasmid DNA. Experiments were performed at least 3 times in triplicate, giving comparable results.

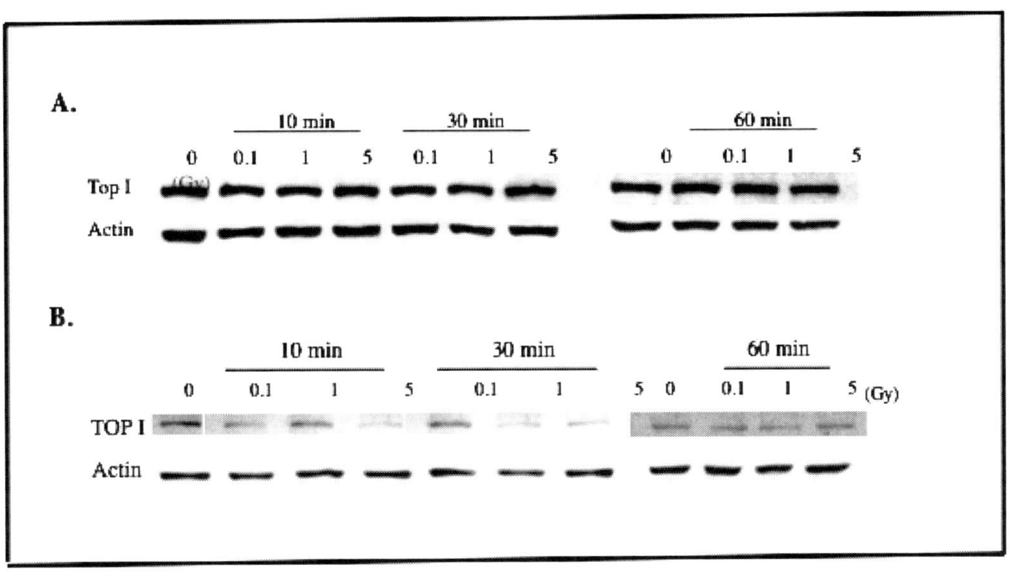

Figure 6. Band depletion-immunoblotting experiments using whole lysates of HCT 116 (A) and HT 29 (B) cells, irradiated with different time point (10, 30 and 60 min of IR exposure) and doses (0.1, 1 and 5 Gy of X ray) and not irradiated (control '0') cells. Experiments were performed twice in triplicate, giving comparable results.

Discussion

Aim of this work was to investigate the role of Top IB in the first onset of repair mechanisms of DNA breaks induced by IR. Despite the fact that Top IB is already used, as target for CPT-derived chemotherapeutic agents, in the treatment of some tumors (Castelli et al., 2012), the underlying mechanisms are not yet fully elucidated (Leppard and Champoux, 2005). The cancer cell lines we used, HCT 116 and HT 29 were chosen because they have been shown to possess the highest levels of Top IB as shows in the NCI-60 cancer cell lines panel (Psister et al., 2009). As reported in literature (Yu and Sinclair, 1967) both cell lines respond to IR with a progressive cell cycle delay with a total block after 15 Gy. Accordingly in both cell lines, 1 hour from IR the amount of SSB increased linearly with dose. Twenty-four hours later the number of SSB strongly decreased indicating an efficient repair capability, in particular in cells exposed to the higher dose. We correlated the kinetic of damage repair with the Top IB activity measured early after exposure to IR. Noteworthy the variation of the Top IB enzyme activity with respect to time are different in the two cell lines. While in HCT 116 cells in vitro activity increased at 30 min from IR and drops at 60 min, in the case of HT 29 activity was early lower and increased later on. The total nuclear amount of enzyme didn't increase after IR, thus we suggest that the higher activity we observed is probably due to post-translational events. The differences in the activation onset showed by the two cell lines, could derive by a delayed timing of the post-translational modifications in HT 29 cells. The first step of Top IB activity consists in its covalent binding with DNA. Shift assay demonstrated that in HT 29 the amount of Top IB bound to DNA correlated with its activity whereas in HCT 116 the binding seemed to vary very little during the period examined. A reason of this discrepancy could be that in HCT 116 the expression level of the enzyme is so large (as evidenced by the rest of the panel) that is not possible to appreciate variations of soluble forms (i.e. the band evaluated in the shift assay). Alternatively, it is possible that the timing of the involvement of the Top IB in the repair process differs in the two cell lines. Remarkably, while the analysis of the enzymatic activity is an index of the level of activation (i.e. of post-translational modifications), the shift-assay is a snapshot of the enzyme state in the living cell (i.e. DNA-bound or free). The amount of MN after 24 hours from IR is higher in HCT 116 comparing to HT 29, unlike from what was seen for the SSB. Anyhow it should be considered that the genetic damages assessed by alkaline Comet assay (base or phosphate alkylation, intermediates in base excision repair, base-less sugars left by glycosidase action) are different from those that cause the formation of MN (DSB, chromosomal loss and mis-repair) (Fenech, 2010). According to Das et al. (2014) following collision with the replication and transcription machineries Top I-linked DNA SSB can be transformed DSB. In HCT 116 Top IB activity was shown to start soon after IR, likely at the peak of the short-lived IR-induced ROS, Top IB catalyzes the unwinding of DNA by creating transient SSB (Gupta et al., 1995). In presence of oxidative stress re-ligation reaction is impaired causing the persistence of DNA breaks (Leppard and Champoux, 2005). ROS also act as mediators of IR-induced MN formation (Choi et al., 2007). The contemporary presence of ROS and the greater number of breaks, may have originated the largest number of MN observed in in the HCT 116 cells. These results are in agreement with our previous finding on two strains of yeast cell *Saccharomyces cerevisiae* (Jahan et al., 2013) showing in Topo + strain a higher early SSB amount comparing to Topo -. Since the increase of the activity of

Top IB is not correlated to the increase of its synthesis, it must be due to a post-translational modification of the protein, therefore the different times of activation / deactivation of the enzyme we observed in the two cell lines are likely due to the different timing of activation of the factors post-translationally modifying the protein probably induced by a kinase activation (Bandyopadhyay and Gjerset, 2011). Phosphorylation is proved to be critical in regulating Top IB activity. Top IB purifies as a phosphoprotein and its ability of binding to DNA is inhibited by treatment with alkaline phosphatase (Pommier Y, 1990). Top IB activity is stimulated in vitro by treatment with the serine kinases, protein kinase C (PKC) or protein kinase CK2 (CK2, formerly casein kinase II) (Samuels et al., 1989; Cardellini et al., 1993; Turman and Douvas et al., 1993). There are many evidences suggesting that one or more Top I serine phosphorylating activities could have a role in regulating Top IB activity affecting the cellular response to CPT-related drugs. However, this possibility has not been fully explored, nor has been identified the critical kinase for achieving an alternative and more effective treatment strategy for curing cancer rather than using Top IB inhibitors. In addition, the knowledge of the mechanisms that intervene immediately to repair the SSB induced by IR and a possible modulation to avoid the formation of DSBs, difficult to repair, could have important application in the prevention of long-term effects of radiation.

REFERENCES

[1] Allan, Y., Chen, S. S., Michael, H., Mace, L., Rothenberg, & Michelle., P. (2004). Induction of Radiosensitization by Indolocarbazole Derivatives: The Role of DNA Topoisomerase I. *Molecular pharmacology*, *66*, 553–560.

[2] Bandyopadhyay, K. & Gjerset, R. A. (2011). Protein kinase CK2 is a central regulator of topoisomerase I hyperphosphorylation and camptothecin sensitivity in cancer cell lines. *Biochemistry*, *50*, 704–714.

[3] Caldecott, K. W. (2003). Protein-protein interactions during mammalian DNA single-strand break repair. *Biochem Soc Trans*, *31*, 247–251.

[4] Cardellini, E. & Durban, E. (1993). Phosphorylation of human topoisomerase I by protein kinase C in vitro and in phorbol 12-myristate 13-acetate-activated HL-60 promyelocytic leukaemia cells. *Biochem J*, *291*, 303–307.

[5] Castelli, S., Campagna, A., Vassallo, O., Tesauro, C., Fiorani, P., Tagliatesta, P., Oteri, F., Falconi, M., Majumder, H. K. & Desideri, A. (2009). Conjugated eicosapentaenoic acid inhibits human topoisomerase IB with a mechanism different from camptothecin. *Arch Biochem Biophys*, *486*, 103–110.

[6] Castelli, S., Coletta, A., D'Annessa, I., Fiorani, P., Tesauro, C. & Desideri, A. (2012). Interaction between natural compounds and human topoisomerase I. *Biol. Chem*, *393*, 1327-1340.

[7] Chen, A. Y., Chou, R., Shih, S. J., Lau, D., Gandara, D. (2004). Enhancement of Radiotherapy with DNA Topoisomerase I-targeted Drugs. *Crit Rev Oncol Hematol*, *50*, 111-119.

[8] Chen, A. Y. & Liu, L. F. (1994). DNA topoisomerases: essential enzymes and lethal targets. *Annu Rev Pharmacol Toxicol*, *34*, 191-218.

[9] Choi, K. M., Kang, C. M., Cho, E. S., Kang, S. M., Lee, S. B., Um H. D. (2007). Ionizing radiation-induced micronucleus formation is mediated by reactive oxygen species that are produced in a manner dependent on mitochondria, Nox1, and JNK. *Oncol rep*, *17*, 1183-1188.

[10] Das, B. B., Huang, S. Y., Murai, J., Rehman, I., Amé, J.C., Sengupta, S., Das, S. K., Majumdar, P., Zhang, H., Biard, D., Majumder, H.K., Schreiber, V. & Pommier, Y. (2014). PARP1-TDP1 coupling for the repair of topoisomerase I-induced DNA damage. *Nucleic Acids Res. 42*, 4435-49. doi: 10.1093/nar/gku088. Epub 2014 Feb 3.

[11] Fenech, M. (2010). The lymphocyte cytokinesis-block micronucleus cytome assay and its application in radiation biodosimetry. *Health Phys*, *98*, 234-43.

[12] Ewesuedo, R. B. & Ratain, M. J. (1997). *Topoisomerase I Inhibitors Oncologist. 23*, 59-364.

[13] García, O., Mandina, T., Lamadrid, A. I., Diaz, A., Remigio, A., Gonzalez, Y., Piloto, J., Gonzalez, J. E. & Alvarez, A. (2004). Sensitivity and variability of visual scoring in the comet assay, results of an inter-laboratory scoring exercise with the use of silver staining. *Mutat. Res*, *556*, 25–34.

[14] Giovanetti, A., Deshpande, T. & Basso, E. (2008). Persistence of genetic damage in mice exposed to low dose of X-rays. *Int. J. Radiat. Biol*, *84*, 227–235.

[15] Gupta, M., Fujimori, A. & Pommier, Y. (1995). Eukaryotic DNA topoisomerases I. *Biochim Biophys Acta*, *1262*, 1–14.

[16] Hall, E. J. & Giaccia A. M. (2006). DNA strand breaks and chromosomal aberrations: In *Radiobiology for the radiologist*, ed. 6th. Philadelphia: Lippincott Williams & Wilkins, New York, 16–29.

[17] Hann, C., Evans, D. L., Fertala, J., Benedetti, P., Bjornsti, M. A. & Hall, D. J. (1998). Increased Camptothecin Toxicity Induced in Mammalian Cells Expressing *Saccharomyces cerevisiae* DNA Topoisomerase I. *J Biol Chem*, *273*, 8425–8433.

[18] Huang, L., Snyder, A. R. & Morgan, W. F. (2003). Radiation-induced genomic instability and its implications for radiation carcinogenesis. *Oncogene*, *22*, 5848-5854.

[19] Husain, I., Mohler, J. L., Seigler, H. F. & Besterman, J. M. (1994). Elevation of topoisomerase I messenger RNA, protein, and catalytic activity in human tumors: demonstration of tumor-type specificity and implications for cancer chemotherapy. *Cancer Res*, *54*, 539–546.

[20] Jahan, Z., Castelli, S., Aversa, G., Rufini, S., Desideri, A. & Giovanetti, A. (2013). Role of human topoisomerase IB on ionizing radiation induced damage. *Biochem Biophys Res Commun*, *432*, 545–548.

[21] Jenkins, G., O'Byrne, K. J., Panizza, B. & Richard, D. J. (2013). Genome Stability Pathways in Head and Neck Cancers. *Int J Genomics*, 2013:464720. doi: 10.1155/2013/464720. Epub 2013 Nov 4.

[22] Leppard, J. B. & Champoux, J. J. (2005). Human DNA topoisomerase I: relaxation, roles, and damage control. *Chromosoma*, *114*, 75-85.

[23] Liehr, I. G., Giovanella B. C., Verschraegen C. F. Eds (2000). *The Camptothecins - unfolding their anticancer potential.*, Ann. N. Y. Acad. Sci. New York,

[24] Lowry, O. H., Rosebrough, N. J., Farr, A. L. & Randall, R. J. (1951). Protein Measurement with the Folin Phenol Reagent. *J. Biol. Chem*, *193*, 265-275.

[25] Mo, Y. Y., Wang, P. & Beck, W. T. (2000). Functional expression of human DNA topoisomerase I and its subcellular localization in HeLa cells. *Exp Cell Res*, *256*, 480–490.

[26] Morgan, W. F., Hartmann, A., Limoli, C. L., Nagar, S. & Ponnaiya, B. (2002). Bystander effects in radiation-induced genomic instability. *Mutat Res*, *504*, 91-100.

[27] Olsen, B. B., Wang, S. Y., Svenstrup, T. H., Chen B. P., Guerra, B. (2012). Protein kinase CK2 localizes to sites of DNA double-strand break regulating the cellular response to DNA damage. *BMC Mol Biol*. 2012 Mar 9;13:7. http://www.biomedcentral.com/1471-2199/13/7.

[28] Pfister, T. D., Reinhold, W. C., Agama, K., Gupta, S., Khin, S. A., Kinders, R. J., Parchment, R. E., Tomaszewski, J. E., Doroshow, J. H. & Pommier, Y. (2009). Topoisomerase I levels in the NCI-60 cancer cell line panel determined by validated ELISA and microarray analysis and correlation with indenoisoquinoline sensitivity. *Mol Cancer Ther*, *8*, 1878–1884.

[29] Pommier, Y., Kerrigan, D., Hartman, K. D. & Glazer, R. I. (1990). Phosphorylation of mammalian DNA topoisomerase I and activation by protein kinase C. *J Biol Chem*, *265*, 9418–9422.

[30] Pommier, Y., Pourquier, P., Urasaki, Y., Wu, J. & Laco, G. S. (1999). Topoisomerase I inhibitors: selectivity and cellular resistance. *Drug Resist Updat*, *2*, 307–318.

[31] Pommier, Y. (2006). Topoisomerase I inhibitors: camptothecins and beyond. *Nature Reviews Cancer*, *6*, 789-802.

[32] Synder, R. D. (1987). Is DNA topoisomerase involved in the UV excision repair process? New evidence from studies with DNA intercalating and non-intercalating anti-tumor agents. *Photochem Photobiol*, *45*, 105–112.

[33] Reinhold, W. C., Mergny, J.L., Liu, H., Ryan, M., Pfister, T. D., Kinders, R., Parchment, R., Doroshow, J., Weinstein, J. N. & Pommier, Y. (2010). Exon array analyses across the NCI-60 reveal potential regulation of TOP1 by transcription pausing at guanosine quartets in the first intron. *Cancer Res*, *70*, 2191–2203.

[34] Saleh-Gohari, N., Bryant, H. E., Schultz N., Parker K. M., Cassel T. & Helleday T. (2005). Spontaneous homologous recombination is induced by collapsed replication forks that are caused by ebdogenous DNA single-strand breaks. *Mol Cell Biol*, *25*, 7158–7169.

[35] Samuels, D. S., Shimizu, Y., Shimizu, N. (1989). Protein kinase C phosphorylates DNA topoisomerase I. FEBS Lett. *259*, 57–60.

[36] St-Amant, C., Lussier, S., Lehoux, J., Laberge, R. M. & Boissonneault, G. (2006). Altered phosphorylation of topoisomerase I following overexpression in an ovarian cancer cell line. *Biochem Cell Biol*, *84*, 55–66.

[37] Stewart L., Redinbo M.R., Qiu X., Hol W.G. & Champoux J. J. (1998). A model for the mechanism of human topoisomerase I, *Science*, *279*, 1534–1541.

[38] Sugimoto, Y., Tsukahara, S., Oh-hara, T., Isoe, T. & Tsuruo, T. (1990). Decreased expression of DNA topoisomerase I in camptothecin-resistant tumor cell lines as determined by a monoclonal antibody. *Cancer Res*, *50*, 6925–6930.

[39] Turman, M. A. & Douvas, A. A. (1993). casein kinase type II (CKII)-like nuclear protein kinase associates with, phosphorylates, and activates topoisomerase I. *Biochem Med Metab Biol*, *50*, 210–225.

[40] Yang, Z. & Champoux, J. (2009). Assays for the preferential binding of human topoisomerase I to supercoiled DNA. *Methods Mol Biol*, *582*, 49–57.

[41] Yu, C., K. & Sinclair, W., K. (1967). Mitotic delay and chromosomal aberrations induced by X-rays in synchronized Chinese hamster cells in vitro. *J Natl Cancer Inst*, *39*, 619–632.

INDEX

D

E

F

G

H

I

J

K

L

M

N

O

P

S

T

U

V

W

X

Y